Computational and Statistical Methods for Analysing Big Data with Applications

Computational and Statistical Methods for Analysing Big Data with Applications

Shen Liu

The School of Mathematical Sciences and the ARC Centre
of Excellence for Mathematical & Statistical Frontiers,
Queensland University of Technology, Australia

James McGree

The School of Mathematical Sciences and the ARC Centre
of Excellence for Mathematical & Statistical Frontiers,
Queensland University of Technology, Australia

Zongyuan Ge

Cyphy Robotics Lab, Queensland University
of Technology, Australia

Yang Xie

The Graduate School of Biomedical Engineering,
the University of New South Wales, Australia

AMSTERDAM • BOSTON • HEIDELBERG • LONDON
NEW YORK • OXFORD • PARIS • SAN DIEGO
SAN FRANCISCO • SINGAPORE • SYDNEY • TOKYO
Academic Press is an imprint of Elsevier

ELSEVIER

Academic Press is an imprint of Elsevier
125, London Wall, EC2Y 5AS.
525 B Street, Suite 1800, San Diego, CA 92101-4495, USA
225 Wyman Street, Waltham, MA 02451, USA
The Boulevard, Langford Lane, Kidlington, Oxford OX5 1GB, UK

Notices
Knowledge and best practice in this field are constantly changing. As new research and experience
broaden our understanding, changes in research methods, professional practices, or medical treatment
may become necessary.

Practitioners and researchers must always rely on their own experience and knowledge in evaluating and
using any information, methods, compounds, or experiments described herein. In using such information
or methods they should be mindful of their own safety and the safety of others, including parties for
whom they have a professional responsibility.

To the fullest extent of the law, neither the Publisher nor the authors, contributors, or editors, assume any
liability for any injury and/or damage to persons or property as a matter of products liability, negligence
or otherwise, or from any use or operation of any methods, products, instructions, or ideas contained in
the material herein.

ISBN: 978-0-12-803732-4

British Library Cataloguing-in-Publication Data
A catalogue record for this book is available from the British Library.

Library of Congress Cataloging-in-Publication Data
A catalog record for this book is available from the Library of Congress.

For Information on all Academic Press publications
visit our website at http://store.elsevier.com/

Working together
to grow libraries in
developing countries

www.elsevier.com • www.bookaid.org

Contents

List of Figures

List of Tables

List of Contributors

C.C. Drovandi School of Mathematical Sciences, Queensland University of Technology, Brisbane, QLD, Australia

Zongyuan Ge Cyphy Robotics Lab, Queensland University of Technology, Australia

C. Holmes Department of Statistics, University of Oxford, Oxford, United Kingdom

Shen Liu The School of Mathematical Sciences and the ARC Centre of Excellence for Mathematical & Statistical Frontiers, Queensland University of Technology, Australia

James McGree The School of Mathematical Sciences and the ARC Centre of Excellence for Mathematical & Statistical Frontiers, Queensland University of Technology, Australia

J.M. McGree School of Mathematical Sciences, Queensland University of Technology, Brisbane, QLD, Australia

K. Mengersen School of Mathematical Sciences, Queensland University of Technology, Brisbane, QLD, Australia

S. Richardson MRC Biostatistics Unit, Cambridge Institute of Public Health, Cambridge, UK

E.G. Ryan School of Mathematical Sciences, Queensland University of Technology, Brisbane, QLD, Australia; Biostatistics Department, Institute of Psychiatry, Psychology and Neuroscience, King's College, London, United Kingdom

Yang Xie The Graduate School of Biomedical Engineering, the University of New South Wales, Australia

Acknowledgment

The authors would like to thank the School of Mathematical Sciences and the ARC Centre of Excellence for Mathematical & Statistical Frontiers at Queensland University of Technology, the Australian Centre for Robotic Vision, and the Graduate School of Biomedical Engineering at the University of New South Wales for their support in the development of this book.

The authors are grateful to Ms Yike Gao for designing the front cover of this book.

Introduction

1

The history of humans storing and analysing data dates back to about 20,000 years ago when tally sticks were used to record and document numbers. Palaeolithic tribespeople used to notch sticks or bones to keep track of supplies or trading activities, while the notches could be compared to carry out calculations, enabling them to make predictions, such as how long their food supplies would last (Marr, 2015). As one can imagine, data storage or analysis in ancient times was very limited. However, after a long journey of evolution, people are now able to collect and process huge amounts of data, such as business transactions, sensor signals, search engine queries, multimedia materials and social network activities. As a 2011 McKinsey report (Manyika et al., 2011) stated, the amount of information in our society has been exploding, and consequently analysing large datasets, which refers to the so-called big data, will become a key basis of competition, underpinning new waves of productivity growth, innovation and consumer surplus.

1.1 What is big data?

Big data is not a new phenomenon, but one that is part of a long evolution of data collection and analysis. Among numerous definitions of big data that have been introduced over the last decade, the one provided by Mayer-Schönberger and Cukier (2013) appears to be most comprehensive:

- Big data is *"the ability of society to harness information in novel ways to produce useful insights or goods and services of significant value"* and *"things one can do at a large scale that cannot be done at a smaller one, to extract new insights or create new forms of value."*

In the community of analytics, it is widely accepted that big data can be conceptualized by the following three dimensions (Laney, 2001):

- Volume
- Velocity
- Variety

1.1.1 Volume

Volume refers to the vast amounts of data being generated and recorded. Despite the fact that big data and large datasets are different concepts, to most people big data implies an enormous volume of numbers, images, videos or text. Nowadays, the amount of information being produced and processed is increasing tremendously, which can be depicted by the following facts:

- 3.4 million emails are sent every second;
- 570 new websites are created every minute;
- More than 3.5 billion search queries are processed by Google every day;
- On Facebook, 30 billion pieces of content are shared every day;
- Every two days we create as much information as we did from the beginning of time until 2003;
- In 2007, the number of Bits of data stored in the digital universe is thought to have exceeded the number of stars in the physical universe;
- The total amount of data being captured and stored by industry doubles every 1.2 years;
- Over 90% of all the data in the world were created in the past 2 years.

As claimed by Laney (2001), increases in data volume are usually handled by utilizing additional online storage. However, the relative value of each data point decreases proportionately as the amount of data increases. As a result, attempts have been made to profile data sources so that redundancies can be identified and eliminated. Moreover, statistical sampling can be performed to reduce the size of the dataset to be analysed.

1.1.2 Velocity

Velocity refers to the pace of data streaming, that is, the speed at which data are generated, recorded and communicated. Laney (2001) stated that the bloom of e-commerce has increased point-of-interaction speed and consequently the pace data used to support interactions. According to the International Data Corporation (https://www.idc.com/), the global annual rate of data production is expected to reach 5.6 zettabytes in the year 2015, which doubles the figure of 2012. It is expected that by 2020 the amount of digital information in existence will have grown to 40 zettabytes. To cope with the high velocity, people need to access, process, comprehend and act on data much faster and more effectively than ever before. The major issue related to the velocity is that data are being generated continuously. Traditionally, the time gap between data collection and data analysis used to be large, whereas in the era of big data this would be problematic as a large portion of data might have been wasted during such a long period. In the presence of high velocity, data collection and analysis need to be carried out as an integrated process. Initially, research interests were directed towards the large-volume characteristic, whereas companies are now investing in big data technologies that allow us to analyse data while they are being generated.

1.1.3 Variety

Variety refers to the heterogeneity of data sources and formats. Since there are numerous ways to collect information, it is now common to encounter various types of data coming from different sources. Before the year 2000, the most common format of data was spreadsheets, where data are structured and neatly fit into tables or relational databases. However, in the 2010s most data are unstructured, extracted from photos, video/audio documents, text documents, sensors, transaction records, etc. The heterogeneity of data sources and formats makes datasets too complex to store and analyse using traditional methods, while significant efforts have to be made to tackle the challenges involved in large variety.

As stated by the Australian Government (http://www.finance.gov.au/sites/default/files/APS-Better-Practice-Guide-for-Big-Data.pdf), traditional data analysis takes a dataset from a data warehouse, which is clean and complete with gaps filled and outliers removed. Analysis is carried out after the data are collected and stored in a storage medium such as an enterprise data warehouse. In contrast, big data analysis uses a wider variety of available data relevant to the analytics problem. The data are usually messy, consisting of different types of structured and unstructured content. There are complex coupling relationships in big data from syntactic, semantic, social, cultural, economic, organizational and other aspects. Rather than interrogating data, those analysing explore it to discover insights and understandings such as relevant data and relationships to explore further.

1.1.4 Another two V's

It is worth noting that in addition to Laney's three Vs, *Veracity* and *Value* have been frequently mentioned in the literature of big data (Marr, 2015). Veracity refers to the trustworthiness of the data, that is, the extent to which data are free of biasedness, noise and abnormality. Efforts should be made to keep data tidy and clean, whereas methods should be developed to prevent recording dirty data. On the other hand, value refers to the amount of useful knowledge that can be extracted from data. Big data can deliver value in a broad range of fields, such as computer vision (Chapter 4 of this book), geosciences (Chapter 5), finance (Chapter 6), civil aviation (Chapter 6), health care (Chapters 7 and 8) and transportation (Chapter 8). In fact, the applications of big data are endless, from which everyone is benefiting as we have entered a data-rich era.

1.2 What is this book about?

Big data involves a collection of techniques that can help in extracting useful information from data. Aiming at this objective, we develop and implement advanced statistical and computational methodologies for use in various high impact areas where big data are being collected.

In Chapter 2, classification methods will be discussed, which have been extensively implemented for analysing big data in various fields such as customer segmentation, fraud detection, computer vision, speech recognition and medical diagnosis. In brief, classification can be viewed as a labelling process for new observations, aiming at determining to which of a set of categories an unlabelled object would belong. Fundamentals of classification will be introduced first, followed by a discussion on several classification methods that have been popular in big data applications, including the k-nearest neighbour algorithm, regression models, Bayesian networks, artificial neural networks and decision trees. Examples will be provided to demonstrate the implementation of these methods.

Whereas classification methods are suitable for assigning an unlabelled observation to one of several existing groups, in practice groups of data may not have been identified. In Chapter 3, three methods for finding groups in data will be introduced: principal component analysis, factor analysis and cluster analysis. Principal component analysis is concerned with explaining the variance-covariance structure of a set of variables through linear combinations of these variables, whose general objectives are data reduction and interpretation. Factor analysis can be considered an extension of principal component analysis, aiming to describe the covariance structure of all observed variables in terms of a few underlying factors. The primary objective of cluster analysis is to categorize objects into homogenous groups, where objects in one group are relatively similar to each other but different from those in other groups. Both hierarchical and non-hierarchical clustering procedures will be discussed and demonstrated by applications. In addition, we will study fuzzy clustering methods which do not assign observations exclusively to only one group. Instead, an individual is allowed to belong to more than one group, with an estimated degree of membership associated with each group.

Chapter 4 will focus on computer vision techniques, which have countless applications in many regions such as medical diagnosis, face recognition or verification system, video camera surveillance, transportation, etc. Over the past decade, computer vision has been proven successful in solving real-life problems. For instance, the registration plate of a vehicle using a tollway is identified from the picture taken by the monitoring camera, and then the corresponding driver will be notified and billed automatically. In this chapter, we will discuss how big data facilitate the development of computer vision technologies, and how these technologies can be applied in big data applications. In particular, we will discuss deep learning algorithms and demonstrate how this state-of-the-art methodology can be applied to solve large scale image recognition problems. A tutorial will be given at the end of this chapter for the purpose of illustration.

Chapter 5 will concentrate on spatial datasets. Spatial datasets are very common in statistical analysis, since in our lives there is a broad range of phenomena that can be described by spatially distributed random variables (e.g. greenhouse gas emission, sea level, etc.). In this chapter, we will propose a computational method for analysing large spatial datasets. An introduction to spatial statistics will be provided at first, followed by a detailed discussion of the proposed computational method. The code of MATLAB programs that are used to implement this method will be listed and discussed next, and a case study of an open-pit mining project will be carried out.

In Chapter 6, experimental design techniques will be considered in analysing big data as a way of extracting relevant information in order to answer specific questions. Such an approach can significantly reduce the size of the dataset to be analysed, and potentially overcome concerns about poor quality due to, for example, sample bias. We will focus on a sequential design approach for extracting informative data. When fitting relatively complex models (e.g. those that are non-linear) the performance of a design in answering specific questions will generally depend upon the assumed model and the corresponding values of the parameters. As such, it is useful to consider prior information for such sequential design problems. We argue that this can be obtained in big data settings through the use of an initial learning phase where data are extracted from the big dataset such that appropriate models for analysis can be explored and prior distributions of parameters can be formed. Given such prior information, sequential design is undertaken as a way of identifying informative data to extract from the big dataset. This approach is demonstrated in an example where there is interest to determine how particular covariates effect the chance of an individual defaulting of their mortgage, and we also explore the appropriateness of a model developed in the literature for the chance of a late arrival in domestic air travel. We will also show that this approach can provide a methodology for identifying gaps in big data which may reveal limitations in the types of inferences that may be drawn.

Chapter 7 will concentrate on big data analysis in the health care industry. Healthcare administrators worldwide are striving to lower the cost of care whilst improving the quality of care given. Hospitalization is the largest component of health expenditure. Therefore, earlier identification of those at higher risk of being hospitalized would help healthcare administrators and health insurers to develop better plans and strategies. In Chapter 7, we will develop a methodology for analysing large-scale health insurance claim data, aiming to predict the number of hospitalization days. Decision trees will be applied to hospital admissions and procedure claims data, which were observed from 242,075 individuals. The proposed methodology performs well in the general population as well as in subpopulations (e.g. elderly people), as the analysis results indicate that it is reasonably accurate in predicting days in hospital.

In Chapter 8 we will study how mobile devices facilitate big data analysis. Two types of mobile devices will be reviewed: wearable sensors and mobile phones. The former is designed with a primary focus on monitoring health conditions of individuals, while the latter are becoming the central computer and communication device in our lives. Data collected by these devices often exhibit high-volume, high-variety and high-velocity characteristics, and hence suitable methods need to be developed to extract useful information from the observed data. In this chapter, we will concentrate on the applications of wearable devices in health monitoring, and a case study in transportation where data collected from mobile devices facilitate the management of road networks.

1.3   Who is the intended readership?

The uniqueness of this book is the fact that it is both methodology- and application-oriented. While the statistical content of this book facilitates analysing big data, the applications and case studies presented in Chapters 4 to 8 can make advanced statistical and computational methods understandable and available for implementation to practitioners in various fields of application.

The primary intended readership of this book includes statisticians, mathematicians or computational scientists who are interested in big-data-related issues, such as statistical methodologies, computing algorithms, applications and case studies. In addition, people from non-statistical/mathematical background but with moderate quantitative analysis knowledge will find our book informative and helpful if they seek solutions to practical problems that are associated with big data. For example, those in the field of computer vision might be interested in image processing and pattern recognition techniques presented in Chapter 4, while the methodology and case study presented in Chapter 7 will be of general interest to practitioners in the health care or medical industry as well as anyone involved in analysing large medical data sets for the purpose of answering specific research questions. Moreover, the case study in Chapter 8, entitled 'mobile devices in transportation', would be beneficial to government officials or industry managers who are responsible for transport projects.

References

Laney, D. (2001). *3D data management: Controlling data volume, velocity, and variety*. US: META Group.

Manyika, J., Chui, M., Brown, B., Bughin, J., Dobbs, R., Roxburgh, C., et al. (2011). *Big data: The next frontier for innovation, competition, and productivity*. US: McKinsey Global Institute.

Marr, B. (2015). *Big data: Using SMART big data, analytics and metrics to make better decisions and improve performance*. UK: Johny Wiley & Sons, Inc.

Mayer-Schönberger, V., & Cukier, K. (2013). *Big data: A revolution that will transform how we live, work, and think*. UK: Hachette.

Classification methods

2

Classification is one of the most popular data mining techniques, which has been extensively used in the analysis of big data in various fields such as customer segmentation, fraud detection, computer vision, speech recognition, medical diagnosis, etc. We consider classification as a labelling process for new observations, that is, it predicts categorical labels for new observations, aiming at determining to which of a set of categories an unlabelled object should belong. For example, a bank loan officer may want to evaluate whether approving a mortgage application is risky or safe, and consequently he/she needs to determine a label for the corresponding applicant ('risky' or 'safe'). This is a typical classification task involving only two categories, whereas in general one may need to classify a new observation into one of k categories where $k = 2, 3,\ldots$. As we will see in the following sections, however, the principles of classification do not alter regardless of the number of classes.

In practice, a classification task is implemented through the following three stages:

Stage 1: Specify a suitable algorithm for classification, that is, a classifier.
Stage 2: Optimize the selected classification algorithm using a set of training data.
Stage 3: Make predictions using the optimized classification algorithm.

The selection of a classifier in Stage 1 is generally data dependent, that is, different studies may favour different classification algorithms. Two issues need to be addressed when selecting a classifier. Firstly, there is a trade-off between the accuracy and computing speed of a classifier: more sophisticated classification algorithms tend to exhibit greater reliability but are associated with larger computation cost. Holding other conditions (e.g. computing power) unchanged, the selection of classifiers heavily depends on what appears to be more desirable to the user: to make predictions quickly or to make prediction accurately. Secondly, the underlying assumptions of a classifier should always be satisfied; otherwise it should not be applied. For instance, one should not employ a linear classifier to solve nonlinear classification problems.

Stage 2 is known as the training process of a classification algorithm, which requires a set of training data which contains observations whose labels are known. The aim of the training process is to derive a rule that can be used to assign each new object to a labelled class optimally. This process is implemented by looking for a set of parameter values that minimizes the likelihood of misclassification.

In Stage 3, the optimized classification algorithm is applied to new observations so that their labels can be predicted. Note that the reliability of predictions builds on the assumption that new observations would behave similarly to those in the training dataset. If this assumption is violated, the performance of the classification algorithm would be questionable.

Computational and Statistical Methods for Analysing Big Data with Applications.
© 2016 Elsevier Ltd. All rights reserved.

In this chapter, the fundamentals of classification are introduced first, followed by a discussion on several classification methods that have been popular in big data applications. Examples will be provided to demonstrate the implementation of these methods.

2.1 Fundamentals of classification

Classification is often referred to as a supervised learning process, that is, all observations in the training dataset are correctly labelled. To describe individual observations, we usually consider a set of features, or explanatory variables, which may be of any data type such as nominal (e.g. gender), ordinal (e.g. ranking), interval (e.g. Celsius temperature) and ratio (e.g. income). By examining the relationship between these features and the corresponding class labels, a classification rule can be derived to discriminate between individuals from different classes.

2.1.1 Features and training samples

We start with the case where $k = 2$, that is, there are two classes of objects. Denote the labels of these two classes π_1 and π_2, respectively. Let R_1 and R_2 be two regions of possible outcomes such that if an observation falls in R_1 it is then categorized into π_1; if it falls in R_2, it is allocated to R_2. When $k = 2$, it is assumed that R_1 and R_2 are collectively exhaustive, that is, R_1 and R_2 jointly cover all possible outcomes. However, one should note that there may not be a clear distinction between R_1 and R_2, implying that they may overlap. As a consequence, classification algorithms are not guaranteed to be error-free. It is possible to incorrectly classify a π_2 object as belonging to π_1 or a π_1 object as belonging to π_2.

Assume that all objects are measured by p random variables, denoted $X' = [X_1, \ldots, X_p]$. It is expected that the observed values of X in π_1 represent the first class, which are different from those in π_2.

For example, consider the following two populations:

π_1: Purchasers of a new product,
π_2: Laggards (those slow to purchase),

whereby consumers are to be categorized into one of these two classes on the basis of observed values of the following four variables, which are presumably relevant in this case:

x_1: Education,
x_2: Income,
x_3: Family size,
x_4: History of brand switching.

That is, an individual object, say the ith object, is characterized by

$$x_i' = \left[x_{1,i}, x_{2,i}, x_{3,i}, x_{4,i}\right].$$

If there are n objects in total, then we have the following data matrix:

$$X = [x_1, \ldots, x_n] = \begin{bmatrix} x_{1,1}, x_{1,2}, \ldots, x_{1,n} \\ x_{2,1}, x_{2,2}, \ldots, x_{2,n} \\ x_{3,1}, x_{3,2}, \ldots, x_{3,n} \\ x_{4,1}, x_{4,2}, \ldots, x_{4,n} \end{bmatrix}.$$

If the n labels of X, corresponding to the n columns, are known, X is considered as a training sample. We then separate X into two groups (purchasers and laggards) and examine the difference to investigate how the label of an object can be determined by its feature values, that is, why the ith consumer is a 'purchaser' or a 'laggard' given the corresponding education, income, family size and history of brand switching.

Example: Discriminating owners from non-owners of riding mowers

In order to identify the best sales prospects for an intensive sales campaign, a riding mower manufacturer is interested in classifying families as prospective owners or non-owners. That is, the following two classes of families are considered:

π_1: Owners of riding mowers,
π_2: Non-owners of riding mowers.

The riding mower manufacturer has decided to characterize each family by two features: income ($1000) and lot size (1000 ft^2). Twelve families are randomly selected from each class and the observed feature values are reported in the following table:

Owners of riding mowers		Non-owners of riding mowers	
Income ($1000)	Lot size (1000 ft^2)	Income ($1000)	Lot size (1000 ft^2)
60	18.4	75	19.6
85.5	16.8	52.8	20.8
64.8	21.6	64.8	17.2
61.5	20.8	43.2	20.4
87	23.6	84	17.6
110.1	19.2	49.2	17.6
108	17.6	59.4	16
82.8	22.4	66	18.4
68	20	47.4	16.4
93	20.8	33	18.8
51	22	51	14
81	20	63	14.8

This dataset is available at http://pages.stat.wisc.edu/~rich/JWMULT02dat/T11-1.DAT.

To learn how the two features can be used to discriminate between owners and non-owners, we display the training sample in the scatterplot below. It can be seen that owners of riding mowers tend to have greater income and bigger lots than non-owners, which implies that if an unlabelled family has a reasonably large income and lot size, it is likely to be predicted as a prospective owner. Nonetheless, one should note from the figure below that the two classes in the training sample exhibit some overlap, indicating that some owners might be incorrectly classified as non-owners and vice versa. The objective of training a classifier is to create a rule that minimizes the occurrence of misclassification. For example, if a linear classifier (represented by the solid line in the figure below) is considered to determine R_1 and R_2, then its parameters (namely, intercept and slope) are optimized such that the number of blue circles falling into R_2 plus the number of green circles falling into R_1 is minimized.

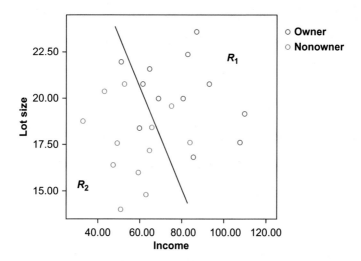

2.1.2   Probabilities of misclassification and the associated costs

In addition to the occurrence of misclassification, one should also consider the cost of misclassification. Generally, the consequence of classifying a π_1 object as belonging to π_2 is different from the consequence of classifying a π_2 object as belonging to π_1. For instance, as we will see in Chapter 8, failing to diagnose a potentially fatal illness is a much more serious mistake than concluding that the disease is present whereas in fact it is not. That is, the cost of the former misclassification is higher than that of the latter. Therefore, an optimal classification algorithm should take into account the costs associated with misclassification.

Denote by $f_1(x)$ and $f_2(x)$ the probability density functions associated with the $p \times 1$ vector random variable X for the populations π_1 and π_2, respectively. Denote $\Omega = R_1 \cup R_2$ the sample space, that is, the collection of all possible observations of X. Two types of misclassification are defined as follows:

$$P(2|1) = P(x \in R_2 | \pi_1) = \int_{R_2} f_1(x)dx,$$
$$P(1|2) = P(x \in R_1 | \pi_2) = \int_{R_1} f_2(x)dx,$$

where $P(2|1)$ refers to the conditional probability that a π_1 object is classified as belonging to π_2, which is computed as the volume formed by the probability density function $f_1(x)$ over the region of π_2. Similarly, $P(1|2)$ denotes the conditional probability of classifying a π_2 object as belonging to π_1, computed as the volume formed by the probability density function $f_2(x)$ over the region of π_1. Furthermore, denote p_1 and p_2 the prior probabilities of π_1 and π_2, respectively, where $p_1 + p_2 = 1$. The overall probabilities of classification can be derived as the product of the prior and conditional probabilities:

- $P(\text{a } \pi_1 \text{ object is correctly classified}) = P(x \in R_1|\pi_1)P(\pi_1) = P(1|1)p_1$,
- $P(\text{a } \pi_1 \text{ object is misclassified}) = P(x \in R_2|\pi_1)P(\pi_1) = P(2|1)p_1$,
- $P(\text{a } \pi_2 \text{ object is correctly classified}) = P(x \in R_2|\pi_2)P(\pi_2) = P(2|2)p_2$,
- $P(\text{a } \pi_2 \text{ object is misclassified}) = P(x \in R_1|\pi_2)P(\pi_2) = P(1|2)p_2$,

and then the associated costs are defined as follows:

- Cost of classifying a π_1 object correctly $= c(1|1) = 0$ (no cost),
- Cost of classifying a π_1 object incorrectly $= c(2|1) > 0$,
- Cost of classifying a π_2 object correctly $= c(2|2) = 0$ (no cost),
- Cost of classifying a π_2 object incorrectly $= c(1|2) > 0$.

The expected cost of misclassification (ECM) is calculated as follows:

$$\text{ECM} = c(2|1)P(2|1)p_1 + c(1|2)P(1|2)p_2,$$

which is considered as the overall cost function to be minimized. Note that if $c(2|1) = c(1|2)$, minimizing the ECM is equivalent to minimizing the total probability of misclassification (TPM), which takes the following form:

$$\text{TPM} = p_1 \int_{R_2} f_1(x)dx + p_2 \int_{R_1} f_2(x)dx.$$

2.1.3 Classification by minimizing the ECM

A reasonable classifier should have the ECM as small as possible, that is, it determines R_1 and R_2 that result in the minimum ECM. Mathematically, R_1 and R_2 are, respectively, defined as the regions over which x satisfies the following inequalities:

$$R_1: \frac{f_1(x)}{f_2(x)} \geq \frac{c(1|2)p_2}{c(2|1)p_1},$$

$$R_2: \frac{f_1(x)}{f_2(x)} < \frac{c(1|2)p_2}{c(2|1)p_1}.$$

That is, an object is classified as π_1 if $f_1(x)c(2|1)p_1 \geq f_2(x)c(1|2)p_2$; otherwise it is classified as π_2.

To implement this classification rule, one needs to evaluate (i) the probability density functions of π_1 and π_2, (ii) the costs of misclassification, and (iii) the prior probabilities of π_1 and π_2. With respect to (i), the probability density functions can be approximated using a training dataset. With respect to (ii) and (iii), it is often much easier to specify the ratios than their component parts. For example, it may be difficult to specify the exact cost of failing to diagnose a potentially fatal illness, but it may be much easier to state that failing to diagnose a potentially fatal illness is, say, ten times as risky as concluding that the disease is present but in fact it is not, that is, the cost ratio is 10.

Example: Medical diagnosis

Suppose that a researcher wants to determine a rule for diagnosing a particular disease based on a collection of features that characterize the symptoms of a patient. Clearly, the following two classes are of interest:

π_1: The disease is present,
π_2: The disease is not present.

The researcher has collected sufficient training data from π_1 and π_2 to approximate $f_1(x)$ and $f_2(x)$, respectively. It is believed that failing to diagnose the disease is five times as risky as incorrectly claiming that the disease is present. In addition, previous studies have found that about 20% of patients would be diagnosed with the disease. Therefore, the following information is available:

$$c(2|1)/c(1|2) = 5/1,$$
$$p_1/p_2 = 0.2/0.8,$$

and then the classification regions are determined as follows:

$$R_1 : \frac{f_1(x)}{f_2(x)} \geq \frac{c(1|2)p_2}{c(2|1)p_1} = \frac{1}{5} \times \frac{0.8}{0.2} = 0.8,$$

$$R_2 : \frac{f_1(x)}{f_2(x)} < \frac{c(1|2)p_2}{c(2|1)p_1} = \frac{1}{5} \times \frac{0.8}{0.2} = 0.8.$$

That is, if $f_1(x)/f_2(x) \geq 0.8$, the corresponding patient is diagnosed with the disease; otherwise the disease is not present.

Assume that a new patient's symptoms are characterized by x_0, with $f_1(x_0) = 0.71$ and $f_2(x_0) = 0.77$. Based on the classification rule above, this patient should be diagnosed with the disease because

$$f_1(x)/f_2(x) = 0.922 > 0.8.$$

Note that under particular circumstances it is nearly impossible to obtain information about the costs of misclassification or prior probabilities. In such cases, it is reasonable to set $c(2|1)/c(1|2)$ and $p_1 = p_2$, implying that the classification rule reduces to

$$R_1 : \frac{f_1(x)}{f_2(x)} \geq 1,$$

$$R_2 : \frac{f_1(x)}{f_2(x)} < 1.$$

In the example above, if $c(2|1)/c(1|2)$ and p_1/p_2 were unknown, the new patient would have been diagnosed as being healthy, since $f_1(x)/f_2(x) = 0.922 < 1$. As a consequence, when carrying out classification tasks one is encouraged to obtain as much information as possible to improve the reliability of classification.

2.1.4 More than two classes

So far we have considered classifying an object into one of two classes, that is, $k = 2$. In general one may need to classify an object into one of k categories $\pi_1, \ldots, \pi_k$ where $k = 2, 3. \ldots$. Suppose that the ith class π_i is associated with the classification region R_i and the probability density function $f_i(x)$, and its prior probability is p_i. For $i, j = 1, \ldots, k$, the probability of classifying a π_i object into π_j is defined as

$$P(j|i) = P(x \in R_j | \pi_i) = \int_{R_j} f_i(x) dx,$$

and the associated cost is denoted $c(j|i)$. The expected cost of misclassifying a π_i object, denoted ECM_i, is computed as

$$\text{ECM}_i = \sum_{j=1}^{k} P(j|i) c(j|i),$$

and the overall ECM is then defined as

$$\text{ECM} = \sum_{i=1}^{k} p_i \text{ECM}_i = \sum_{i=1}^{k} p_i \sum_{j=1}^{k} P(j|i) c(j|i).$$

By minimizing the ECM, the following classification rule applies to a new observation x_0:

- x_0 is classified to π_j if $\sum_{i=1, i \neq j}^{k} p_i f_i(x_0) c(j|i) \leq \sum_{i=1, i \neq l}^{k} p_i f_i(x_0) c(l|i)$ for any $l \neq j$, where $j, l = 1, \ldots, k$.

If $c(j|i)$ is a constant for any $i \neq j$, the classification rule is then dominated by the posterior probabilities. For a new observation x_0, its posterior probability associated with π_i is of the following form:

$$P(\pi_i|x_0) = P(x_0|\pi_i)P(\pi_i)/P(x_0),$$

where $P(x_0|\pi_i) = f_i(x_0)$, $P(\pi_i) = p_i$, and $P(x_0) = \sum_{i=1}^{k} P(x_0|\pi_i)P(\pi_i) = \sum_{i=1}^{k} f_i(x_0)p_i$. Since $P(x_0)$ remains the same regardless of the i value, the following classification rule applies:

- x_0 is allocated to π_j if $f_j(x_0)p_j \geq f_i(x_0)p_i$ for any $i \neq j$, where $i,j \in \{1, \ldots, k\}$.

Now we have learnt the fundamentals of classification. In the next section, we will discuss several classifiers that have been widely employed for analysing big data.

2.2 Popular classifiers for analysing big data

The following classifiers have been found popular for analysing big data:

- k-Nearest neighbour algorithm,
- Regression models,
- Bayesian networks,
- Artificial neural networks,
- Decision trees.

In this section, we briefly discuss these classifiers (as a comprehensive introduction is out of the scope of this book). For each classifier, suitable references are provided to which the interested reader is referred for full details.

2.2.1 k-Nearest neighbour algorithm

Being a non-parametric, instance-based method, the k-nearest neighbour (kNN) algorithm is arguably the simplest and most intuitively appealing classification procedure (Hall, Park, & Samworth, 2008). Given a new observation x_0, the kNN algorithm determines its class label as follows:

- Within the training sample, select k data points ($k = 1, 2, \ldots$) that are closest to x_0, which are referred to as the k-nearest neighbours. The closeness is evaluated by some distance measure, for example, Euclidean distance.
- The class label of x_0 is determined by a majority vote of its k-nearest neighbours, that is, x_0 is assigned to the class that is most frequent among its k-nearest neighbours.

To demonstrate, consider the 'German credit' dataset which is available at the following address:

https://archive.ics.uci.edu/ml/datasets/Statlog + %28German + Credit + Data%29

This dataset consists of records of 1000 customers characterized by 20 features, with the response variable being whether a customer is a good or bad credit risk. For the sake of simplicity, we only consider two features: duration in months and credit amount. The 1000 records are split into a training set and a test set, where the former contains the first 800 customers and the latter contains the rest. The following MATLAB code is used to implement the kNN algorithm:

```
y = creditrisk; % Response variable: good or Bad credit risk
x = [duration, credit_amount]; % Duration in months and Credit amount

train = 800; % Size of training sample
xtrain = x(1:train,:); % Training sample
ytrain = y(1:train,:); % Labels of training sample

mdl = fitcknn(xtrain,ytrain,'NumNeighbors',5); % Train the kNN with k = 5

xtest = x(train + 1:end,:); % Test set
ytest = y(train + 1:end,:); % Labels of test set

ypredict = predict(mdl,xtest); % Prediction of the test set
rate = sum(ypredict == ytest)/numel(ytest); % Compute the rate of correct
classification
```

The following table summarizes the classification results:

		Predicted	
		Good	Bad
Actual	Good	123	16
	Bad	47	14

Results show that 68.5% of the customers in the test set were correctly labelled. This rate may be enhanced by considering a different set of explanatory variables (features), or a different value of k. Hall et al. (2008) provided guidelines for the selection of k. When applying kNN to large datasets, it is computationally intensive to conduct exhaustive searching, that is, the distance from x_0 to each training sample is computed. To make kNN computationally tractable, various nearest-neighbour search algorithms have been proposed to reduce the number of distance evaluations actually performed (e.g. Nene & Nayar, 1997). In addition, feature extraction and dimension reduction can be performed to avoid the effects of the curse of dimensionality, as in high dimensions the Euclidean distance is not helpful in the evaluation of closeness.

Although the kNN algorithm is conceptually very simple, it can perform very well in numerous real-life applications. For instance, as we will see in Chapter 8, the kNN algorithm is capable of discriminating between the biosignal of a fall and those of daily-living activities (e.g. walking, jumping, etc.), which enables automatic fall detection.

2.2.2 Regression models

A regression model is used to estimate the relationship between a response (dependent) variable and a set of explanatory (independent) variables, which indicates the strength and significance of impact of the explanatory variables on the response variable. Generally, a regression model can be expressed in the following form:

$$Y = f(X) + \varepsilon,$$

where Y denotes the response variable, X is the collection of explanatory variables, and ε is known as the error term.

Although there are innumerable forms of regression models, for the purpose of classification we are interested in those which are capable of predicting categorical response variables. If the response variable is binary, the logistic regression can be considered (Long, 1997). Denote Y the response variable which has two possible outcomes (0 or 1), and $X = [X_1, \ldots, X_p]'$ denotes the p explanatory variables that may influence Y. The aim of the logistic regression is to estimate the following conditional probability:

$$P(Y = 1|X),$$

that is, the probability that Y is predicted as '1' given values of the p explanatory variables. To ensure that $P(Y = 1|X)$ takes values between zero and one, the logistic function is considered:

$$\sigma(t) = 1/(1 + \exp(-t)),$$

where $t \in (-\infty, +\infty)$ and $\sigma(t) \in [0, 1]$. Based on the $\sigma(t)$ function, the logistic regression model is formulated as follows:

$$P(Y = 1|X) = \frac{\exp(\beta_0 + \beta_1 X_1 + \cdots + \beta_p X_p)}{1 + \exp(\beta_0 + \beta_1 X_1 + \cdots + \beta_p X_p)},$$

where β_0 denotes the intercept and $\beta_1, \ldots, \beta_p$ are the p coefficients associated with $X_1, \ldots, X_p$, respectively. The $p + 1$ parameters are usually estimated by the maximum likelihood estimator, and then $P(Y = 1|x_0)$ can be computed where x_0 denotes a new observation. If $P(Y = 1|x_0)$ is greater than τ where τ denotes a threshold value, the new observation is classified into the first category ($Y = 1$); otherwise it is assigned to the second group ($Y = 0$). If no other information is available (i.e. costs of misclassification and prior probabilities), $\tau = 0.5$ is a reasonable choice.

To illustrate, let's revisit the example: discriminating owners from non-owners of riding mowers. The logistic regression takes the following form:

$$P(Y = 1|X) = \frac{\exp(\beta_0 + \beta_1 X_1 + \beta_2 X_2)}{1 + \exp(\beta_0 + \beta_1 X_1 + \beta_2 X_2)},$$

where $Y = 1$ stands for the group of owners, X_1 and X_2 denote income and lot size, respectively. Using the dataset discussed in Section 2.1.1, the following parameter estimates are obtained: $\hat{\beta}_0 = -25.86$, $\hat{\beta}_1 = 0.110$ and $\hat{\beta}_2 = 0.963$. Now suppose we have the following new observations:

	Income	Lot size
Person 1	90	22
Person 2	64	20
Person 3	70	18

Then based on the estimated regression model and the 0.5 threshold, it is predicted that

	Probability of being an owner	Classification result
Person 1	0.9947	Owner
Person 2	0.6089	Owner
Person 3	0.3054	Non-owner

If Y has more than two classes, one may consider a multinomial logistic regression model. Suppose that $Y \in \{1, \ldots, k\}$, that is, it has k possible outcomes. Given a collection of explanatory variables $X = [X_1, \ldots, X_p]'$, the multinomial logistic regression is of the following form (Long & Freese, 2001):

$$P(Y = m|X) = \frac{\exp(\beta_{0,m|b} + \beta_{1,m|b}X_1 + \cdots + \beta_{p,m|b}X_p)}{\sum_i^k \exp(\beta_{0,i|b} + \beta_{1,i|b}X_1 + \cdots + \beta_{p,i|b}X_p)},$$

where b denotes the base category, and $\beta_{0,i|b}, \ldots, \beta_{p,i|b}$ are the coefficients that measure the ith category in relation to the base category. That is, the multinomial logistic regression consists of $k - 1$ equations with $k - 1$ sets of parameters, each of which evaluates the likelihood of a particular outcome in the comparison to a pre-determined benchmark, namely, the base category. Note that $\beta_{0,b|b}, \ldots, \beta_{p,b|b} = 0$.

Moreover, to determine an optimal set of $X_1, \ldots, X_p$, a variable selection algorithm can be implemented. The following MATLAB function is designed to choose from all possible regression models (i.e. all possible combinations of explanatory variables):

```
function [M, Iset] = allregress(y,X,criteria)

% Inputs:
% y - dependent variable, column vector, n-by-1
% X - n-by-p explanatory variable matrix
% criteria - model selection criteria
%

% Outputs:
% M - Statistics of the selected model
```

```
% Iset - indices of the selected variables, a logic row vector, 1-by-k
%
% Example:
% y = randn(100,1);
% X = randn(100,6);
% [M_BIC, I_BIC] = allregress(y,X);
% [M_AIC, I_AIC] = allregress(y,X,'AIC');

if size(y,1) ~ = size(X,1)
       error('y and X must have the same length')
end

if nargin < = 2
       criteria = 'BIC'; % Set default criterion
end

n = size(y,1); % No. of observations
k = size(X,2); % No. of explanatory variables
nsets = 2^k-1; % No. of all possible regression models

C = NaN(1); % used to store criterion values
Iset = NaN(1,k); % used to store index of X variables

for i = 1 : nsets
       iset = logical(dec2bin(i, k) - '0');
       x = X(:,iset);
       [~,~,M] = glmfit(x,y,'binomial','link','logit');
       MSE = mean(M.resid.^2);
       switch criteria
         case 'AIC'
           c = log(MSE) + numel(M.beta)*2/n;
         case 'BIC'
           c = log(MSE) + numel(M.beta)*log(n)/n;
         case 'HQ'
           c = log(MSE) + numel(M.beta)*2*log(log(n))/n;
         otherwise
           error('Selection criterion is not defined properly')
       end
       if isnan(C)
         C = c;
         Iset = iset;
       elseif C > c
         C = c;
         Iset = iset;
       end
end
M = regstats(y,X(:,Iset));
end
```

2.2.3 Bayesian networks

A Bayesian network is a probabilistic graphical model that measures the conditional dependence structure of a set of random variables based on the Bayes theorem:

$$P(A|B) = \frac{P(B|A)P(A)}{P(B)}.$$

Pearl (1988) stated that Bayesian networks are graphical models that contain information about causal probability relationships between variables and are often used to aid in decision making. The causal probability relationships in a Bayesian network can be suggested by experts or updated using the Bayes theorem and new data being collected. The inter-variable dependence structure is represented by nodes (which depict the variables) and directed arcs (which depict the conditional relationships) in the form of a directed acyclic graph (DAG). As an example, the following DAG indicates that the incidence of two waterborne diseases (diarrhoea and typhoid) depends on three indicators of water samples: total nitrogen, fat/oil and bacteria count, each of which is influenced by another layer of nodes: elevation, flooding, population density and land use. Furthermore, flooding may be influenced by two factors: percolation and rainfall.

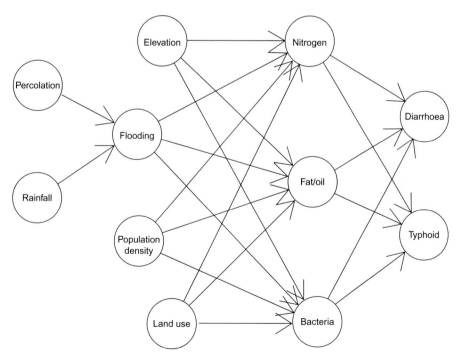

There are two components involved in learning a Bayesian network: (i) structure learning, which involves discovering the DAG that best describes the causal relationships in the data, and (ii) parameter learning, which involves learning about the conditional probability distributions. The two most popular methods for determining the structure of the DAG are the DAG search algorithm (Chickering, 2002) and the K2 algorithm (Cooper & Herskovits, 1992). Both of these algorithms assign equal prior probabilities to all DAG structures and search for the structure that maximizes the probability of the data given the DAG, that is, $P(\text{data}|\text{DAG})$ is maximized. This probability is known as the Bayesian score. Once the DAG structure is determined, the maximum likelihood estimator is employed as the parameter learning method. Note that it is often critical to incorporate prior knowledge about causal structures in the parameter learning process. For instance, consider the causal relationship between two binary variables: rainfall (large or small) and flooding (whether there is a flood or not). Apparently, the former has impact on the latter. Denote the four corresponding events:

     R: Large rainfall,
     $\overline{R}$: Small rainfall,
     F: There is a flood,
     $\overline{F}$: There is no flood.

In the absence of prior knowledge, the four joint probabilities $P(F, R)$, $P(\overline{F}, R)$, $P(F, \overline{R})$ and $P(\overline{F}, \overline{R})$ need to be inferred using the observed data; otherwise, these probabilities can be pre-determined before fitting the Bayesian network to data. Assume that the following statement is made by an expert: if the rainfall is large, the chance of flooding is 60%; if the rainfall is small, the chance of no flooding is four times as big as that of flooding. Then we have the following causal relationship as prior information:

$$P(F|R) = 0.6,$$
$$P(\overline{F}|R) = 0.4,$$
$$P(F|\overline{R}) = 0.2,$$
$$P(\overline{F}|\overline{R}) = 0.8.$$

Furthermore, assume that meteorological data show that the chance of large rainfalls is 30%, namely, $P(R) = 0.3$. Then the following contingency table is determined:

	R	$\overline{R}$
F	$P(F,R) = P(F|R)P(R) = 0.18$	$P(F,\overline{R}) = P(F|\overline{R})P(\overline{R}) = 0.14$
$\overline{F}$	$P(\overline{F},R) = P(\overline{F}|R)P(R) = 0.12$	$P(\overline{F},\overline{R}) = P(\overline{F}|\overline{R})P(\overline{R}) = 0.56$

which is an example of pre-specified probabilities based on prior knowledge.

For the purpose of classification, the naïve Bayes classifier has been extensively applied in various fields, such as the classification of text documents (spam or legitimate email, sports or politics news, etc.) and automatic medical diagnosis (to be introduced in Chapter 8 of this book). Denote y the response variable which has

k possible outcomes, that is, $y \in \{y_1, \ldots, y_k\}$, and let $x_1, \ldots, x_p$ be the p features that characterize y. Using the Bayes theorem, the conditional probability of each outcome, given $x_1, \ldots, x_p$, is of the following form:

$$P(y_i|x_1, \ldots, x_p) = \frac{P(x_1, \ldots, x_p|y_i)P(y_i)}{P(x_1, \ldots, x_p)}.$$

Note that the naïve Bayes classifier assumes that $x_1, \ldots, x_p$ are mutually independent. As a result, $P(y_i|x_1, \ldots, x_p)$ can be re-expressed as follows:

$$P(y_i|x_i, \ldots, x_p) = \frac{P(y_i)\prod_{j=1}^{p}P(x_j|y_i)}{P(x_1, \ldots, x_p)},$$

which is proportional to $P(y_i)\prod_{j=1}^{p}P(x_j|y_i)$. The maximum *a posteriori* decision rule is applied and the most probable label of y is determined as follows:

$$\hat{y} = \mathrm{argmax}_{i \in \{1, \ldots, k\}}P(y_i) \prod_{j=1}^{p} P(x_j|y_i).$$

An illustration of the naïve Bayes classifier is provided here. Let's revisit the 'German credit' dataset. This time, we consider the following features: duration in months, credit amount, gender, number of people being liable, and whether a telephone is registered under the customer's name. The setting of training and test sets remains the same. The following MATLAB code is implemented:

```
y = creditrisk; % Response variable
x = [duration, credit_amount, male, nppl, tele]; % Features

train = 800; % Size of training sample

xtrain = x(1:train,:); % Training sample
ytrain = y(1:train,:); % Labels of training sample

xtest = x(train+1:end,:); % Test set
ytest = y(train+1:end,:); % Labels of test set

nbayes = fitNaiveBayes(xtrain,ytrain); % Train the Naïve Bayes

ypredict = nbayes.predict(xtest); % Prediction of the test set
rate = sum(ypredict == ytest)/numel(ytest); % Compute the rate of correct
classification
```

Again, 68.5% of the customers in the test set were correctly labelled. The results are summarized in the following table:

		Predicted	
		Good	Bad
Actual	Good	121	18
	Bad	45	16

For modelling time series data, dynamic Bayesian networks can be employed to evaluate the relationship among variables at adjacent time steps (Ghahramani, 2001). A dynamic Bayesian network assumes that an event has impact on another in the future but not vice versa, implying that directed arcs should flow forward in time. A simplified form of dynamic Bayesian networks is known as hidden Markov models. Denote the observation at time t by Y_t, where t is the integer-valued time index. As stated by Ghahramani (2001), the name 'hidden Markov' is originated from two assumptions: (i) a hidden Markov model assumes that Y_t was generated by some process whose state S_t is hidden from the observer, and (ii) the states of this hidden process satisfy the first-order Markov property, where the rth order Markov property refers to the situation that given $S_{t-1}, \ldots, S_{t-r}$, S_t is independent of S_τ for $\tau < t - r$. The first-order Markov property also applies to Y_t with respect to the states, that is, given S_t, Y_t is independent of the states and observations at all other time indices. The following figure visualizes the hidden Markov model:

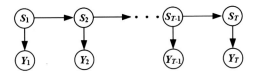

Mathematically, the causality of a sequence of states and observations can be expressed as follows:

$$P(Y_1, S_1, \ldots, Y_T, S_T) = P(S_1)P(Y_1|S_1) \prod_{t=2}^{T} P(S_t|S_{t-1})P(Y_t|S_t).$$

Hidden Markov models have shown potential in a wide range of data mining applications, including digital forensics, speech recognition, robotics and bioinformatics. The interested reader is referred to Bishop (2006) and Ghahramani (2001) for a comprehensive discussion of hidden Markov models.

In summary, Bayesian networks appear to be a powerful method for combining information from different sources with varying degrees of reliability. More details of Bayesian networks and their applications can be found in Pearl (1988) and Neapolitan (1990).

2.2.4 Artificial neural networks

The development of artificial neural networks (ANNs), or simply neural networks, is inspired by biological neural networks (e.g. the neural network of a human brain), which are used to approximate functions based on a large number of inputs. The basic unit of an ANN is known as perceptron (Rosenblatt 1985), which is a single-layer neural network. The operation of perceptrons is based on error-correction learning (Haykin, 2008), aiming to determine whether an input belongs to one class or another (Freund & Schapire, 1999). Given an input vector $x = [+1, x_1, \ldots, x_m]'$, a perceptron produces a single binary output y, optimizing the corresponding weights $w = [b, w_1, \ldots, w_m]'$. The architecture of a perceptron is represented by the following figure (Haykin, 2008):

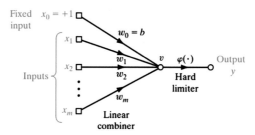

The linear combiner output is of the following form:

$$v = \sum_{i=0}^{m} w_i x_i = w'x,$$

where $w_0 = b$ is known as the bias term. The actual response y is determined as follows:

$$y = \mathrm{sgn}\left(w'x\right),$$

where $\mathrm{sgn}\left(w'x\right) = +1$ if $w'x > 0$ and -1 if otherwise.

To optimize the weight vector w, Haykin (2008) designed the following algorithm:

- If the nth member of the training set, denoted $x(n)$, is correctly classified by the weight vector $w(n)$ that is computed at the nth iteration, no correction is made to the weight vector in the next iteration, namely, $w(n + 1) = w(n)$.
- If $x(n)$ is incorrectly classified by $w(n)$, then the weight vector is updated as follows:

$$w(n + 1) = w(n) - \eta(n)x(n),$$

where $\eta(n)$ denotes the learning-rate parameter which controls the adjustment applied to the weight vector at the nth iteration. Although $\eta(n)$ may change from one iteration to the next, it is convenient to set $\eta(n) = \eta$, which is a constant, which leads to a fixed-increment adaptation rule for the perceptron. Haykin (2008) proved that for a fixed learning rate, the perceptron converges after n_0 iterations in the sense that $w(n_0) = w(n_0 + 1) = w(n_0 + 2) = \cdots$.

An alternative way to adapt the weight vector was proposed by Lippmann (1987), known as the perceptron convergence algorithm:

1. Set $w(0) = [0, 0, \ldots, 0]'$.
2. At the nth iteration, compute the actual response $y(n) = \mathrm{sgn}(w'(n)x(n))$.

3. Apply the following adaptation rule: $w(n + 1) = w(n) + \eta(d(n) - y(n))x(n)$, where $0 < \eta \leq 1$ and $d(n)$ denotes the desired response. $d(n) - y(n)$ can be considered as an error signal. Note that if $x(n)$ is correctly classified, $d(n) = y(n)$, implying $w(n + 1) = w(n)$.
4. Set $n = n + 1$ and go back to Step 2.

The well-known support vector machines can be interpreted as an extension of the perceptron. A support vector machine classifies objects by constructing a hyperplane that separates different groups, where the hyperplane is determined such that the distance from it to the nearest training data point of any class is maximized (Vapnik, 1998). If the classes of original data points are not linearly separable, kernel functions are usually considered to map data points into a higher-dimensional space so that the separability of data groups can be improved. More details of support vector machines will be provided in Section 4.2.2 of this book.

The learning algorithms of perceptrons can be applied to ANNs, since an ANN can be decomposed into a collection of single-layer neural networks. Consider the following ANN:

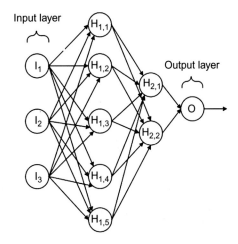

where the leftmost and rightmost layers are known as the input and output layers, respectively, while the intermediate layers are referred to as hidden layers. It can be seen that each node in the output and hidden layers is connected to all neurons in the preceding layer, exhibiting the same architecture as a single-layer perceptron. Nonetheless, training an ANN is much more complicated than training a perceptron, since one would need to optimize a much larger number of parameters (weights). In Chapter 4 of this book, we will further investigate how an ANN can be constructed and trained by a sample of data. As a branch of ANNs, deep neural networks will be introduced in Chapter 4, which have been proven very powerful as a means of classification.

2.2.5 Decision trees

A decision tree is a predictive model, which uses a tree-like graph to map the observed data of an object to conclusions about the target value of this object.

The decision tree is known as a classification tree if the target variable takes a finite set of values, whereas it is referred to as a regression tree if the target variable is continuous. The leaf nodes of a classification tree correspond to class labels, while its branches represent conjunctions of features that lead to those class labels.

To explain how a decision tree can be constructed, consider the following training sample:

$$X = [x_1, \ldots, x_n] = \begin{bmatrix} x_{1,1} & \cdots & x_{1,n} \\ \vdots & \ddots & \vdots \\ x_{p,1} & \cdots & x_{p,n} \end{bmatrix},$$

where n denotes the number of observations and p is the number of features that characterize $x_1, \ldots, x_n$. A decision tree can be determined by the following algorithm:

1. At the beginning, $x_1, \ldots, x_n$ are considered as a single group.
2. Set j_1, where $j_1 \in \{1, \ldots, p\}$;
    for i in $1:n$
    if $x_{j_1,i} \geq c_{j_1}$ where c_{j_1} denotes some threshold of the j_1th feature
        x_i is categorized into Subgroup 1;
    else
        x_i is categorized into Subgroup 2;
    end
    end
3. Set j_2, where $j_2 \in \{1, \ldots, p\}$ and $j_2 \neq j_1$;
    for i in g_1 where g_1 denotes the indices of objects in Subgroup 1
    if $x_{j_2,i} \geq c_{j_2}$ where c_{j_2} denotes some threshold of the j_2th feature
        x_i is categorized into Sub-subgroup 1-1;
    else
        x_i is categorized into Sub-subgroup 1-2;
    end
    end
    Set j_3, where $j_3 \in \{1, \ldots, p\}$ and $j_3 \neq j_1$;
    for i in g_2 where g_2 denotes the indices of objects in Subgroup 2
    if $x_{j_3,i} \geq c_{j_3}$ where c_{j_3} denotes some threshold of the j_3th feature
        x_i is categorized into Sub-subgroup 2-1;
    else
        x_i is categorized into Sub-subgroup 2-2;
    end
    end
4. The process continues until some stopping criterion is satisfied.

During the training process, $j_1, j_2, j_3, \ldots, c_{j_1}, c_{j_2}, c_{j_3}, \ldots$ are optimized so that the incidence of misclassification is minimized. Note that the algorithm above carries out a binary splitting at each node, whereas in practice a group can be split into more than two subgroups. Once a decision tree is grown, the prediction of class labels can be made by following the decisions in the tree from the root node down to a leaf node, while each leaf node corresponds to a class label.

Once more, let's consider the 'German credit' dataset for the purpose of illustration. The following four categorical features are employed to simplify the decision tree: gender, number of people being liable, whether a telephone is registered under the customer's name, and whether the customer is a foreign worker. The setting of training and test sets remains the same. The following MATLAB code is implemented to construct a decision tree:

```
y = creditrisk; % Response variable
x = [male, nppl, tele, foreign]; % Features

train = 800; % Size of training sample

xtrain = x(1:train,:); % Training sample
ytrain = y(1:train,:); % Labels of training sample

xtest = x(train+1:end,:); % Test set
ytest = y(train+1:end,:); % Labels of test set

ctree = fitctree(xtrain,ytrain); % Create decision tree
```

Both text description and graphic description of the constructed decision tree are available:

```
view(ctree) % text description
view(ctree,'mode','graph') % graphic description
```

Output:

```
Decision tree for classification
1  if x1 < 0.5 then node 2 elseif x1 > = 0.5 then node 3 else 1
2  if x3 < 0.5 then node 4 elseif x3 > = 0.5 then node 5 else 1
3  class = 1
4  if x2 < 1.5 then node 6 elseif x2 > = 1.5 then node 7 else 1
5  class = 1
6  class = 1
7  class = 2
```

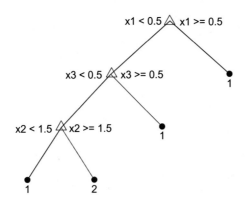

Then the constructed decision tree is used for prediction:

```
ypredict = predict(ctree,xtest);
rate = sum(ypredict == ytest)/numel(ytest);
```

Sixty nine percentage of the customers in the test set were correctly labelled. The results are summarized in the following table:

		Predicted	
		Good	Bad
Actual	Good	138	1
	Bad	61	0

It should be stressed that although the 69% accuracy is slightly higher than the outcome in the previous sections (68.5%), the classification result seems unreliable. As one can observe from the table above, all the customers that are actually "bad" were predicted as "good". The main reason is that the cost of misclassification was not taken into account, implying that the severity of predicting a "bad" customer as "good" was treated equally as that of predicting a "good" customer as "bad". Apparently, the former mistake is more serious than the latter in the context of banking and finance, and consequently the cost of the former should be higher than that of the latter. In MATLAB, the cost of misclassification may be specified using a square matrix, where the (i, j) element denotes the cost of classifying an individual into class j if its true class is i. Incorporating the cost matrix in the decision tree algorithm is left as an exercise for the reader.

Ensemble techniques are often applied to decision trees in order to enhance the performance of classification. One of them is named random forests, which represents a collection of decision trees (tree predictors) such that each tree depends on the values of a random vector sampled independently and with the same distribution for all trees in the forest (Breiman, 2001). In other words, random forests attempt to build multiple decision trees with different data samples and initial variables and then aggregate predictions to make a final decision. Suppose we have n observations which are characterized by p features. The following algorithm describes how random forests can be implemented:

1. Decide on the size of sub-samples, denoted by n^* where $n^* \le n$. Decide on the number of features that are considered for the construction of decision trees, denoted p^* where $p^* \le p$.
2. Create a bootstrap sample of size n^*, that is, randomly take n^* data points out of the original n observations with replacement.
3. Build a decision tree for the bootstrap sample. At each node, p^* features are selected at random out of the original p features, and then the best split of the node is determined based on the selected p^* features. The splitting process continues until the decision tree is grown to the largest extent.
4. Repeat Steps 2 and 3 many times, and then the predictions from different decision trees are integrated to make an aggregated prediction.

One may use the R package 'randomForest' to carry out classification using random forests:

https://cran.r-project.org/web/packages/randomForest/index.html

Breiman (2001) claimed that random forests are an effective tool in classification, since injecting the right kind of randomness makes them accurate classifiers. One advantage of random forests is that they are powerful in processing large datasets with high dimensionality, being able to identify the most significant variables out of thousands of inputs. It is for this reason that random forests have been very popular in big data analytics.

2.3 Summary

The classification methods discussed in this chapter are suitable for assigning an unlabelled observation to one of several existing groups, whereas in practical situations groups of data may not have been identified, meaning that class labels of training data are not known *a priori*. In the next chapter, approaches to finding groups in data will be introduced.

References

Bishop, C. M. (2006). *Pattern recognition and machine learning*. New York, NY: Springer.
Breiman, L. (2001). Random forests. *Machine Learning*, *45*, 5−32.
Chickering, D. M. (2002). Optimal structure identification with greedy search. *Journal of Machine Learning Research*, *3*, 507−554.
Cooper, G. F., & Herskovits, E. (1992). A Bayesian method for the induction of probabilistic networks from data. *Machine Learning*, *9*, 309−347.
Freund, Y., & Schapire, R. E. (1999). Large margin classification using the perceptron algorithm. *Machine Learning*, *37*, 277−296.
Ghahramani, Z. (2001). An introduction to hidden Markov models and Bayesian networks. *International Journal of Pattern Recognition and Artificial Intelligence*, *15*, 9−42.
Hall, P., Park, B. U., & Samworth, R. J. (2008). Choice of neighbor order in nearest-neighbor classification. *The Annals of Statistics*, *36*, 2135−2152.
Haykin, S. O. (2008). *Neural networks and learning machines* (3rd ed.). Upper Saddle River, NJ: Prentice Hall.
Lippmann, R. P. (1987). An introduction to computing with neural nets. *IEEE ASSP Magazine*, *4*, 4−22.
Long, J. S. (1997). *Regression models for categorical and limited dependent variables*. Thousand Oaks, CA: Sage Publications, Inc.
Long, J. S., & Freese, J. (2001). *Regression models for categorical dependent variables using stata*. College Station, TX: Stata Press.
Neapolitan, R. E. (1990). *Probabilistic reasoning in expert systems*. New York, NY: John Wiley & Sons, Inc.
Nene, S. A., & Nayar, S. K. (1997). A simple algorithm for nearest neighbor search in high dimensions. *IEEE Transactions on Pattern Analysis and Machine Intelligence*, *19*, 989−1003.
Pearl, J. (1988). *Probabilistic reasoning in intelligent systems: Networks of plausible inference*. San Mateo, CA: Morgan Kaufmann Publishers, Inc.
Rosenblatt, F. (1985). The perceptron: A probabilistic model for information storage and organization in the brain. *Psychological Review*, *6*, 386−408.
Vapnik, V. (1998). *Statistical learning theory*. New York, NY: John Wiley and Sons, Inc.

Finding groups in data

3

Consider the following data matrix:

$$\begin{pmatrix} x_{1,1} & \cdots & x_{1,p} \\ \vdots & \ddots & \vdots \\ x_{n,1} & \cdots & x_{n,p} \end{pmatrix},$$

that is, there are n observations in the dataset, each of which is characterized by p variables. If n or p is considerably large, finding groups of x's is often advantageous to data analysts. For example, in medical research patients are characterized by numerous variables such as age, gender, height, weight, medical history, etc., whereas the number of patients under study can be very large. If groups of these patients can be identified, subsequent processing of data would be easier since one does not need to treat individuals in the same group differently.

Three statistical methods are discussed in this chapter: principal component analysis (PCA), factor analysis (FA) and cluster analysis (CA). PCA and FA aim to examine the inter-relationship among characterizing variables, while the objective of CA is to investigate the similarity/dissimilarity among observations.

PCA is concerned with explaining the variance-covariance structure of a set of variables through linear combinations of these variables for the general objective of data reduction and interpretation. It can reveal relationships that were not previously suspected and thereby can facilitate new insights and understanding of data. PCA frequently serves as intermediate steps in much larger investigations. For example, principal components may be used as inputs into a multiple regression.

On the other hand, FA can be considered an extension of PCA. The purpose of FA is to describe the covariance structure of all observed variables in terms of a few underlying factors, which are unobserved, random quantities. The assumption behind FA is that variables can be grouped by correlations, that is, those within a particular group are highly correlated themselves but have much weaker correlations with those in a different group. The aim of FA is to identify a factor for each of those groups. For example, correlations from the group of test scores in mathematics, physics, chemistry and biology may suggest an underlying "intelligence" factor. A second group of variables, representing physical-fitness scores, might correspond to another factor. It is this type of structure that FA seeks to confirm.

Cluster Analysis is a technique that produces groups (clusters) of data. It is therefore often used in segmentation studies to identify, for example, market segments. Inter-relationships among the whole set of observations are examined. The primary objective

Computational and Statistical Methods for Analysing Big Data with Applications.

of cluster analysis is to categorize objects into homogenous groups based on the set of variables considered. Objects in a group are relatively similar in terms of these variables and different from objects in other groups. The clustering is carried out on the basis of a measure of similarity, and the produced clusters are comprised of data points which are most similar.

3.1 Principal component analysis

Suppose X consists of p variables:

$$X = [X_1, X_2, \ldots, X_p]',$$

where $X_i = [x_{1,i}, \ldots, x_{n,i}]$, $i = 1, \ldots, p$. Initially, all p variables are needed to reproduce the variability of X. However, in most cases a large portion of such variability can be captured by k variables, where $k < p$. These k variables are known as principal components, each of which is a linear combination of $X_1, X_2, \ldots, X_p$. In other words, the information that can be provided by the k principal components is almost as much as that in the original p variables, and therefore X can be grouped by replacing $X_1, X_2, \ldots, X_p$ by the k principal components.

It would be helpful to consider the geometrical meaning of principal components. Since the k principal components are linear combinations of the p variables, geometrically it implies that a new coordinate system is constructed by rotating the original system whose axes are $X_1, X_2, \ldots, X_p$. The axes of the new system represent directions of maximum variability, enabling a more parsimonious description of the variability of data.

Principal components depend solely on the covariance matrix of data. Denote Σ the covariance matrix of X, with eigenvalues $\lambda_1 \geq \lambda_2 \geq \ldots \geq \lambda_p \geq 0$ and the associated eigenvectors $e_1, e_2, \ldots, e_p$, respectively. The ith principal component, sometimes referred to as the ith factor, is of the following form:

$$Y_i = e_{i,1}X_1 + e_{i,2}X_2 + \cdots + e_{i,p}X_p,$$

where $i = 1, \ldots, p$ and $e_{i,1}, \ldots, e_{i,p}$ are the p elements of the ith eigenvector e_i. The following properties hold:

$$Var(Y_i) = e'_i \Sigma e_i = \lambda_i; \tag{3.1}$$

$$Cov(Y_i, Y_j) = e'_i \Sigma e_j = 0 \text{ for } i \neq j; \tag{3.2}$$

$$\sum_{i=1}^{p} Var(X_i) = \sum_{i=1}^{p} Var(Y_i) = \lambda_1 + \lambda_2 + \cdots + \lambda_p. \tag{3.3}$$

Equation (3.1) implies that the variances of principal components are respectively equal to the associated eigenvalues, and Eqn (3.2) indicates that these components are uncorrelated. The third property shows that the total variance of X is equal to the total variance of principal components, which is equal to the sum of the p eigenvalues.

Note that the second and third properties above jointly imply that the principal components are the "decorrelated" representatives of the original data, as the total variance is fully accounted for. The proportion of total variance explained by the jth principal component is $\lambda_j / \sum \lambda_i$, which evaluates the importance of Y_j. Such proportions are useful when evaluating the joint importance of a number of principal components. For example, if $(\lambda_1 + \lambda_2 + \lambda_3) / \sum \lambda_i = 0.8$, it means the first three principal components Y_1, Y_2 and Y_3 jointly account for 80% of the total variability in X.

This leads to a critical question: how many principal components do we need to describe the covariance structure of data sufficiently? Obviously it is possible to compute as many principal components as there are variables, but no parsimony is gained in doing so. For data grouping purposes, a smaller number of components should be extracted. Several procedures have been suggested for determining the number of principal components, which are discussed briefly as follows (refer to Reddy and Acharyulu (2009) for details):

- A *priori* determination. This applies when the researcher knows how many principal components to expect and therefore k can be specified beforehand. For example, although there are many variables describing the performance of a basketball player, the coach would expect three main factors: offence, defence and teamwork.
- Determination based on eigenvalues. An eigenvalue represents the amount of variance associated with the factor. Hence, only factors with a variance greater than 1.0 are included. Factors with variance less than 1.0 are no better than a single variable, as each variable would have a variance of 1.0 after standardization. This approach will result in a conservative number of factors when $p < 20$.
- Determination based on scree plots. A scree plot displays eigenvalues against the number of principal components in order of extraction, and its shape indicates an appropriate k value. Typically, the plot has a distinct break between the steep slope associated with large eigenvalues and a gradual trailing off associated with the rest. This gradual trailing off is referred to as the scree, and the point at which it starts denotes the appropriate k value.
- Determination based on percentage of variance. The k value is determined so that the proportion of variance explained jointly by the k principal components reaches a satisfactory level. What level of variance is satisfactory depends upon the problem, but it is recommended that at least 60% of the total variance should be accounted for.

Once the model is fitted, factor scores associated with each principal component can be calculated. For the purpose of data summarization, computing factor scores is of critical importance, since these scores will be used instead of original variables in subsequent analysis. That is, the original $n \times p$ data matrix is summarized by an $n \times k$ matrix, where $k < p$. Most computer programs have PCA functions that allow users to request factor scores.

In practice, one should follow the four steps below when carrying out PCA:

1. Determine the appropriateness of PCA;
2. Decide on the number of principal components;
3. Estimate the loadings of principal components;
4. Calculate factor scores.

To illustrate how the PCA is implemented, the dataset "Evaluation.csv" is considered. It consists of course evaluation data for a university in the United States to predict the overall rating of lectures based on ratings of teaching skills, instructor's knowledge of the material, the expected grade and some other variables. For the sake of simplicity, only 12 variables are employed, which are Item 13−Item 24 in the dataset. These were responses to questions where students had to state how strongly they agreed with a particular statement.

The first task is to obtain the correlation matrix of those 12 variables. For PCA to be appropriate, the variables must be correlated. It is expected that highly correlated variables would also highly correlate with the same principal component. In the correlation matrix, there are some very high correlations, and many of the others are greater than 0.3.

row.names	Item 13	Item 14	Item 15	Item 16	Item 17	Item 18	Item 19	Item 20	Item 21	Item 22	Item 23	Item 24
Item 13	1.0000000	0.6601990	0.5648586	0.5323979	0.5203005	0.3788634	0.2628427	0.3069115	0.4447557	0.3101425	0.5309872	0.4302788
Item 14	0.6601990	1.0000000	0.6264745	0.4897919	0.5343690	0.4143193	0.3082334	0.3089306	0.4193585	0.3109440	0.5373871	0.4309684
Item 15	0.5648586	0.6264745	1.0000000	0.5134614	0.5707953	0.4618413	0.3658660	0.3491391	0.4782028	0.3527890	0.5738384	0.4344244
Item 16	0.5323979	0.4897919	0.5134614	1.0000000	0.5735621	0.3951156	0.3180128	0.3283784	0.4206244	0.3305147	0.4539613	0.4174130
Item 17	0.5203005	0.5343690	0.5707953	0.5735621	1.0000000	0.5663240	0.4523730	0.3863314	0.5677362	0.4308977	0.6025244	0.4987991
Item 18	0.3788634	0.4143193	0.4618413	0.3951156	0.5663240	1.0000000	0.6306455	0.4569862	0.5393199	0.5090651	0.5589572	0.4550560
Item 19	0.2628427	0.3082334	0.3658660	0.3180128	0.4523730	0.6306455	1.0000000	0.4095614	0.4776591	0.4628939	0.4330585	0.3560574
Item 20	0.3069115	0.3089306	0.3491391	0.3283784	0.3863314	0.4569862	0.4095614	1.0000000	0.3895163	0.3753599	0.3984233	0.3560242
Item 21	0.4447557	0.4193585	0.4782028	0.4206244	0.5677362	0.5393199	0.4776591	0.3895163	1.0000000	0.4877615	0.6050334	0.4813750
Item 22	0.3101425	0.3109440	0.3527890	0.3305147	0.4308977	0.5090651	0.4628939	0.3753599	0.4877615	1.0000000	0.4922145	0.4348084
Item 23	0.5309872	0.5373871	0.5738384	0.4539613	0.6025244	0.5589572	0.4330585	0.3984233	0.6050334	0.4922145	1.0000000	0.6827084
Item 24	0.4302788	0.4309684	0.4344244	0.4174130	0.4987991	0.4550560	0.3560574	0.3560242	0.4813750	0.4348084	0.6827084	1.0000000

To determine whether PCA could prove useful for grouping variables, two methods can be applied. The first is called the Bartlett's test of sphericity (Bartlett, 1937), which is used to test the null hypothesis that the variables are uncorrelated, that is, Σ is a diagonal matrix. Its test statistic follows a Chi-square distribution with $p(p-1)/2$ degrees of freedom. Rejecting the null hypothesis means PCA is appropriate. The second method is called the Kaiser-Meyer-Olkin (KMO) measure of sampling adequacy (Kaiser, 1970). This index compares the magnitudes of the observed correlation coefficients to the magnitudes of the partial correlation coefficients, which is used to determine the appropriateness of PCA as a means of data reduction. Small values of KMO imply that correlations between pairs of variables cannot be explained well by other variables and therefore PCA may not be appropriate. In contrast, high values (between 0.5 and 1.0) indicate that the observed data have a simpler and more interpretable representation and hence PCA would be useful as a means of data reduction.

The Bartlett's test statistic returned 8569.12 with a p-value $= 0.000$, suggesting that there is strong evidence against the null hypothesis. The KMO measure was computed to be 0.930, indicating that PCA is appropriate. The next step is to determine how many principal components would be sufficient to represent the covariance structure. The following scree plot is produced, which indicates $k = 2$ would be reasonable. This is supported by the eigenvalues, as only two of them are greater than 1.0.

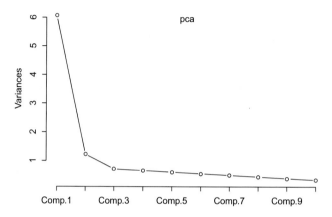

In Step 3, the coefficients of principal components are estimated. The importance of components is evaluated. It can be seen that the first principal component explained 50.6% of the total variation, while the second accounted for 10.2%. The first two principal components are responsible for 60.8% of the total variation in the data. The loadings of all 12 principal components are estimated. Note that the blank entries in the matrix of loadings are small but not zero. These loadings are associated with unrotated factors, which are typically difficult to interpret. Usually, one would consider rotating the factors to make the interpretation easier. The rotation of factors will be discussed in the next section.

```
> summary(pca)
Importance of components:
                          Comp.1    Comp.2     Comp.3     Comp.4     Comp.5     Comp.6     Comp.7     Comp.8     Comp.9    Comp.10
Standard deviation     2.4643447 1.1057958 0.85730686 0.8050843 0.75656695 0.73428701 0.69667683 0.6547363 0.61920783 0.57767483
Proportion of Variance 0.5060829 0.1018987 0.06124792 0.0540134 0.04769946 0.04493145 0.04044655 0.0357233 0.03195153 0.02780902
Cumulative Proportion  0.5060829 0.6079816 0.66922954 0.7232429 0.77094239 0.81587385 0.85632040 0.8920437 0.92399522 0.95180424
                         Comp.11    Comp.12
Standard deviation     0.55769261 0.51703780
Proportion of Variance 0.02591842 0.02227734
Cumulative Proportion  0.97772266 1.00000000
> loadings(pca)

Loadings:
        Comp.1 Comp.2 Comp.3 Comp.4 Comp.5 Comp.6 Comp.7 Comp.8 Comp.9 Comp.10 Comp.11 Comp.12
Item 13 -0.285 -0.421                      0.567                -0.171 -0.536  -0.117
Item 14 -0.290 -0.396 -0.133         0.389  0.174  0.148         0.309  0.649
Item 15 -0.303 -0.271 -0.150         0.244        -0.711 -0.336 -0.171 -0.227  -0.207
Item 16 -0.275 -0.265 -0.225        -0.769         0.191        -0.271 -0.177   0.201   0.152
Item 17 -0.323                0.176 -0.272 -0.226 -0.138 -0.205  0.624  0.473  -0.227
Item 18 -0.303  0.321 -0.162  0.214  0.144 -0.167  0.167         0.408 -0.658   0.197  -0.126
Item 19 -0.257  0.446 -0.294  0.348  0.160 -0.189  0.265  0.231 -0.413  0.363  -0.187
Item 20 -0.235  0.270 -0.430 -0.814               -0.129
Item 21 -0.302  0.144  0.164  0.152               -0.817  0.174 -0.236         0.237  -0.154
Item 22 -0.259  0.350  0.229        -0.179  0.820  0.145 -0.142
Item 23 -0.331         0.403         0.150 -0.161                -0.103 -0.124   0.799
Item 24 -0.286         0.600 -0.301        -0.304  0.341        -0.128                 -0.476

        Comp.1 Comp.2 Comp.3 Comp.4 Comp.5 Comp.6 Comp.7 Comp.8 Comp.9 Comp.10 Comp.11 Comp.12
SS loadings      1.000  1.000  1.000  1.000  1.000  1.000  1.000  1.000  1.000  1.000   1.000   1.000
Proportion Var   0.083  0.083  0.083  0.083  0.083  0.083  0.083  0.083  0.083  0.083   0.083   0.083
Cumulative Var   0.083  0.167  0.250  0.333  0.417  0.500  0.583  0.667  0.750  0.833   0.917   1.000
```

3.2 Factor analysis

Factor analysis seeks to answer the question: does the factor model, with a relatively small number of factors, adequately represent the observed data?

Recall $X = [X_1, X_2, \ldots, X_p]'$ with covariance matrix Σ. The factor model assumes that X is linearly dependent on m common factors $F_1, \ldots, F_m$ and p error terms $\varepsilon_1, \ldots, \varepsilon_p$:

$$X_i = \mu_i + l_{i,1}F_1 + l_{i,2}F_2 + \cdots + l_{i,m}F_m + \varepsilon_i,$$

where $i = 1, \ldots, p$ and μ_i is the mean of X_i. The coefficient $l_{i,j}$ is known as the loading of the ith variable X_i on the jth factor F_j. Note that the $m + p$ random variables $F_1, \ldots, F_m, \varepsilon_1, \ldots, \varepsilon_p$ that are used to express X are unobserved. This distinguishes factor analysis from regression analysis where explanatory variables are observed.

Let $F = [F_1, \ldots, F_m]'$, $\varepsilon = [\varepsilon_1, \ldots, \varepsilon_p]'$ and

$$L = \begin{bmatrix} l_{1,1} & \cdots & l_{1,m} \\ \vdots & \ddots & \vdots \\ l_{p,1} & \cdots & l_{p,m} \end{bmatrix},$$

the following results are of importance:

F and ε are independent;

$E(F) = 0$, $Cov(F) = I$ where I denotes the identity matrix;

$E(\varepsilon) = 0$, $Cov(\varepsilon) = \psi$ where ψ is a diagonal matrix.

$$Cov(X) = LL' + \psi; \tag{3.4}$$

Equation (3.4) implies that $Var(X_i) = \sum_{j=1}^{m} l_{i,j}^2 + \psi_i$ and $Cov(X_i, X_k) = \sum_{j=1}^{m} l_{i,j} l_{k,j}$, which specifies the covariance structure of the observed data in terms of factor loadings. It should be stressed that these results are a consequence of the assumption that the factor model is linear in common factors. If the p variables have nonlinear relationship with $F_1, \ldots, F_m$, the covariance structure represented by $LL' + \psi$ might be inadequate.

The estimation of L, F and ε is based on the sample covariance matrix, denoted S. If S deviates significantly from a diagonal matrix, then factor analysis is worth doing. There are two popular methods for parameter estimation, namely, the principal component method and the maximum likelihood method. It is possible that these two methods produce distinct estimates in particular cases. Note also that if the off-diagonals of S are small, then the variables do not have a desirable strength of correlation. In these cases the factor analysis is not worth carrying out, as the specific variances or uniqueness play the dominant role. To investigate statistically whether the factor analysis is worth doing or not, the same two procedures described in the previous section can be applied, namely, the Bartlett's test and the Kaiser-Meyer-Olkin (KMO) measure.

The communality of the ith variable, known as the ith communality, is defined as the portion of its variance that is explained by common factors, which is equal to $\sum_{j=1}^{m} l_{i,j}^2$. In contrast, the portion of the ith variable's variance due to ε_i is called the specific variance, or uniqueness, which is ψ_i. The factor model assumes that all elements in Σ can be reproduced from the $p \times m$ elements of L and the p diagonal elements of ψ. If $p = m$, Σ can be reproduced exactly by LL' with $\psi = 0$, but in this case groups of data are not produced. It is when $m \ll p$ that factor analysis is most useful. Deciding on the value of m is exactly the same as deciding on the value of k, and therefore those procedures described before can be applied. After an appropriate m value is selected, the communalities are computed, providing information about the proportion of each variable's variance that is accounted for by the m common factors. The sums of squared loadings are the variances associated with the m factors, and the percentage of variance explained by a single factor can be computed accordingly.

An important output from factor analysis is the factor matrix consisting of (unrotated) factor loadings. Factor loadings indicate correlations between factors and variables. It is possible that particular factors are correlated with many variables, and therefore the factor loading matrix is not easy to interpret. In such cases, the interpretation of factor loadings is often enhanced by rotating factors. Ideally, after rotation each factor will have large loadings for only some of the variables, and each variable will have large loadings with only a few factors. Rotation does not have impact on communalities, but the percentage of variance accounted for by each factor does change because the variance explained by individual factors is redistributed after rotation. The VARIMAX procedure is widely applied as a means of orthogonal rotation, which minimizes the number of variables with high loadings on a factor. Following this procedure, the rotated factors are uncorrelated. It is worth noting that allowing for inter-factor correlations may sometimes simplify the interpretation, especially when factors in the population tend to be strongly correlated.

Same as before, factor scores can be computed as a means of data reduction. Furthermore, the adequacy of the fitted model can be assessed by investigating the correlation matrix computed from the loadings of common factors:

$$\hat{\rho} = L_R L'_R,$$

where L_R denotes the matrix of rotated factor loadings. $\hat{\rho}$ is then compared to the original correlation matrix. If the off-diagonal elements of the two matrices are reasonably close, representing the $n \times p$ data matrix by $n \times m$ factor scores is satisfactory.

In practice, six tasks need to be performed when carrying out factor analysis:

1. Determine the appropriateness of factor analysis (same as that in PCA);
2. Decide on the number of common factors (same as that in PCA);
3. Estimate the model;

4. Interpret (rotated) factor loadings;
5. Compute factor scores;
6. Evaluate the adequacy of the fitted model.

To demonstrate, the dataset "Evaluation.csv" is revisited here. It has been shown in the previous section that factor analysis is worth carrying out, and $m = 2$ appears to be a reasonable choice. A model with two common factors is then estimated using the maximum-likelihood technique.

The communalities show the proportion of variance of each variable that is accounted for by the factor model, based on two factors. Ideally, these values should be close to 1. For most variables, a reasonable proportion of variance has been explained by the factor model.

```
> communalities
   Item13    Item14    Item15    Item16    Item17    Item18    Item19    Item20    Item21    Item22    Item23    Item24
0.6333212 0.6372355 0.5763726 0.4445413 0.5948879 0.6400044 0.5263493 0.3132888 0.5291612 0.4299620 0.6288660 0.4463940
```

The rotated factor loadings are reported, which are easier to interpret than unrotated factor loadings. The interpretation is facilitated by identifying variables that have large loadings on the same factor, and then that factor can be interpreted in terms of the identified variables.

```
> fit <- factanal(data, 2, rotation="varimax") # Rotated factors
> fit$loadings

Loadings:
        Factor1 Factor2
Item13  0.200   0.770
Item14  0.228   0.765
Item15  0.348   0.675
Item16  0.317   0.587
Item17  0.527   0.563
Item18  0.746   0.289
Item19  0.709   0.152
Item20  0.499   0.253
Item21  0.602   0.408
Item22  0.613   0.233
Item23  0.566   0.556
Item24  0.491   0.453

               Factor1 Factor2
SS loadings      3.200   3.200
Proportion Var   0.267   0.267
Cumulative Var   0.267   0.533
```

The first common factor has relatively high loadings on [17] instructor uses clear examples, [18] instructor is sensitive to students, [19] instructor allows me to ask questions, [20] instructor is accessible to students outside class, [21] instructor is aware of students' understanding, [22] I am satisfied with student performance evaluation, [23] compared to other instructors and [24] compared to other courses. The second common factor exhibits high loadings on [13] instructor is well prepared, [14] instructor scholarly grasp, [15] instructor confidence, [16] instructor focus, [17] instructor uses clear examples, [21] instructor aware of student understanding,

[23] compared to other instructors and [24] compared to other courses. It appears that the first factor is mainly about how the instructor interacts with students, and hence it may be interpreted as an underlying nature or approachability measure. The second factor seems to describe how well the instructor knows and presents the course, and therefore it may be interpreted as a measure of instructor's competency or performance. Overall, the loadings reveal that higher lecture ratings are associated with instructors being more competent and approachable.

A two-column matrix consisting of factor scores is then obtained. As there are too many rows in the matrix, it is not reported here. Based on the rotated factor loadings, the correlation matrix is reproduced, and then the model performance is assessed.

```
> round(rhat,2) # Display reproduced correlation matrix
        Item13 Item14 Item15 Item16 Item17 Item18 Item19 Item20 Item21 Item22 Item23 Item24
Item13   0.63   0.63   0.59   0.52   0.54   0.37   0.26   0.29   0.43   0.30   0.54   0.45
Item14   0.63   0.64   0.60   0.52   0.55   0.39   0.28   0.31   0.45   0.32   0.55   0.46
Item15   0.59   0.60   0.58   0.51   0.56   0.45   0.35   0.34   0.49   0.37   0.57   0.48
Item16   0.52   0.52   0.51   0.44   0.50   0.41   0.31   0.31   0.43   0.33   0.51   0.42
Item17   0.54   0.55   0.56   0.50   0.59   0.56   0.46   0.41   0.55   0.45   0.61   0.51
Item18   0.37   0.39   0.45   0.41   0.56   0.64   0.57   0.45   0.57   0.52   0.58   0.50
Item19   0.26   0.28   0.35   0.31   0.46   0.57   0.53   0.39   0.49   0.47   0.49   0.42
Item20   0.29   0.31   0.34   0.31   0.41   0.45   0.39   0.31   0.40   0.36   0.42   0.36
Item21   0.43   0.45   0.49   0.43   0.55   0.57   0.49   0.40   0.53   0.46   0.57   0.48
Item22   0.30   0.32   0.37   0.33   0.45   0.52   0.47   0.36   0.46   0.43   0.48   0.41
Item23   0.54   0.55   0.57   0.51   0.61   0.58   0.49   0.42   0.57   0.48   0.63   0.53
Item24   0.45   0.46   0.48   0.42   0.51   0.50   0.42   0.36   0.48   0.41   0.53   0.45
> round(resid,2) # Display the difference between original and reproduced version
        Item13 Item14 Item15 Item16 Item17 Item18 Item19 Item20 Item21 Item22 Item23 Item24
Item13   0.37   0.03  -0.02   0.02  -0.02   0.01   0.00   0.01   0.01   0.01  -0.01  -0.02
Item14   0.03   0.36   0.03  -0.03  -0.02   0.02   0.03   0.00  -0.03  -0.01  -0.02  -0.03
Item15  -0.02   0.03   0.42   0.01   0.01   0.01   0.02   0.00  -0.01  -0.02   0.00  -0.04
Item16   0.02  -0.03   0.01   0.56   0.08  -0.01   0.00   0.02  -0.01   0.00  -0.05   0.00
Item17  -0.02  -0.02   0.01   0.08   0.41   0.01  -0.01  -0.02   0.02  -0.02  -0.01  -0.02
Item18   0.01   0.02   0.01  -0.01   0.01   0.36   0.06   0.01  -0.03  -0.02  -0.02  -0.04
Item19   0.00   0.03   0.02   0.00  -0.01   0.06   0.47   0.02  -0.01  -0.01  -0.05  -0.06
Item20   0.01   0.00   0.00   0.02  -0.02   0.01   0.02   0.69  -0.01   0.01  -0.02   0.00
Item21   0.01  -0.03  -0.01  -0.01   0.02  -0.03  -0.01  -0.01   0.47   0.02   0.04   0.00
Item22   0.01  -0.01  -0.02   0.00  -0.02  -0.02  -0.01   0.01   0.02   0.57   0.02   0.03
Item23  -0.01  -0.02   0.00  -0.05  -0.01  -0.02  -0.05  -0.02   0.04   0.02   0.37   0.15
Item24  -0.02  -0.03  -0.04   0.00  -0.02  -0.04  -0.06   0.00   0.00   0.03   0.15   0.55
```

Based on the fitted factor model, the correlation matrix of data can be reproduced by the loadings of the two common factors. The comparison of the reproduced correlation matrix to the original version evaluates the factor model. It can be seen that most of the reproduced correlations are similar to the observed correlations, with the greatest difference being around 0.15. Most differences are less than 0.05. It is then concluded that for the observed dataset, reducing from $p = 12$ to $m = 2$ appears to be satisfactory.

3.3 Cluster analysis

Cluster analysis can be used as a general data processing tool to identify clusters or subgroups of data that are more manageable than individual observations. Subsequent multivariate analysis is conducted on the clusters rather than on the individual observations. For example, instead of focusing on each consumer, a marketing manager would find it much more efficient to conduct customer behaviour analysis of groups of individuals. Finding groups of customers, also known as customer segmentation, is commonly considered by e-commerce companies like eBay and Amazon to maximize profit. In fact, cluster analysis is one of the most widely applied statistical methods in big data analytics.

3.3.1 Hierarchical clustering procedures

Clustering procedures can be either hierarchical or nonhierarchical. Hierarchical clustering is characterized by the development of a hierarchy or tree like structure, in either an agglomerative or a divisive manner. Agglomerative hierarchical methods are commonly used. These methods start with individual variables, that is, at the beginning there are as many clusters as there are items. Most similar items are grouped together first, and these groups are merged further according to similar characteristics, until finally all subgroups are merged into one cluster. The process can be summarized as follows:

1. Start with N clusters, each containing a single variable, and therefore an N by N symmetric matrix of distances (or similarities);
2. Search the distance matrix for the nearest (which means most similar) pair of clusters. Let this distance between the "most similar" clusters U and V be denoted by d_{UV}. Remember that for this first step, the distance is between two variables as each unit is a cluster at this point in time;
3. Merge clusters U and V. This will form a new cluster, labelled UV. The entries in the distance matrix must then be updated by deleting the rows and columns corresponding to cluster U and V and adding a row and column giving the distances between cluster UV and the remaining clusters;
4. Repeat steps 2 and 3 a total of $N - 1$ times. This will result in all variables being in a single cluster at the end of the algorithm. At each step, record the identity of clusters that are merged and the distances at which the merger takes place.

Note that in Step 2 and 3, a function is required to evaluate the between-cluster distance. For this purpose, linkage and variance methods can be used. If a linkage method is considered, the average linkage (average distance) is usually preferred to its competitors, namely, the single linkage (minimum distance or nearest neighbour) and the complete linkage (maximum distance or farthest neighbour), as it utilizes information on all pairwise distances. On the other hand, the variance methods attempt to generate clusters which minimize the within-cluster variance. A commonly used variance method is the Ward's procedure (Ward, 1963). For each cluster, the means for all the variables are computed, and then the sum of squared distance to the cluster mean is calculated. At each stage, the two clusters with the smallest increase in the overall sum of squares within cluster distances are combined.

Of the hierarchical clustering methods, the average linkage and Ward's methods have been shown to perform better than the other procedures. To demonstrate, three groups of bivariate Gaussian realizations are simulated based on the following parameter settings:

Group 1: $\mu_1 = \begin{bmatrix} 2 \\ 2 \end{bmatrix}$, $\Sigma_1 = \begin{bmatrix} 2 & 1.5 \\ 1.5 & 3 \end{bmatrix}$;

Group 2: $\mu_2 = \begin{bmatrix} 2 \\ 8 \end{bmatrix}$, $\Sigma_2 = \begin{bmatrix} 3 & 2 \\ 2 & 4 \end{bmatrix}$;

Group 3: $\mu_3 = \begin{bmatrix} 8 \\ 8 \end{bmatrix}$, $\Sigma_3 = \begin{bmatrix} 4 & 0 \\ 0 & 3 \end{bmatrix}$.

The size of each group is 1000. The following plot visualizes the simulated 3000 individuals:

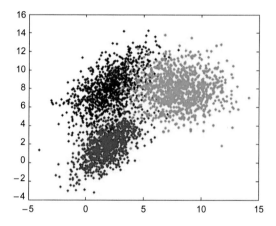

The single, complete, average linkage and Ward's methods are applied to determine groups of the simulated individuals, where the Euclidean distance is considered to measure the dissimilarity. The results are displayed in the following figure:

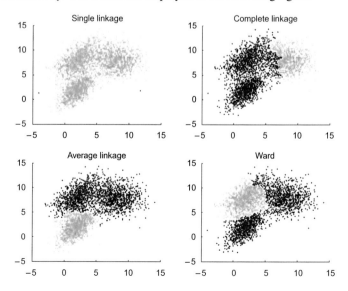

It is observed from this figure that the average linkage and Ward's methods outperformed the two competitors, as the resulting clusters are fairly similar to the simulated groups. In contrast, the single linkage method performed poorly, whereas the complete linkage method could not separate Group 1 and 2 well. The following table reports the clustering accuracy of each method, that is, the proportion of individuals that were grouped correctly:

	Single	Complete	Average	Ward's
Accuracy	16.8%	79.4%	90.1%	91.1%

The MATLAB code used for this demonstration is provided in the appendix of this chapter.

3.3.2 Nonhierarchical clustering procedures

A nonhierarchical clustering method produces a single solution by grouping items into a collection of k clusters, where k is a prespecified number. Unlike hierarchical algorithms, nonhierarchical procedures do not produce a tree like clustering structure, and the pairwise distances of all individuals in the dataset are not required. It is for this reason that nonhierarchical clustering procedures are much more efficient for large datasets.

To initiate the clustering procedure, cluster seeds, which represent the starting points of clusters, need to be specified beforehand. Choosing these seeds should be free of bias, and therefore a reliable way is to choose the starting points randomly.

The k-means algorithm proposed by MacQueen (1967) is probably the most widely applied nonhierarchical clustering procedure. A brief description is provided here. Let $X = \{X_1, X_2, \ldots, X_n\}$ be a set of n objects, where $X_i = (x_{i,1}, x_{i,2}, \ldots, x_{i,m})$, characterized by m variables. The k-means algorithm partitions X into k clusters by minimizing the objective function P with unknown variables U and Z:

$$P(U, Z) = \sum_{l=1}^{k} \sum_{i=1}^{n} \sum_{j=1}^{m} u_{i,l} d(x_{i,j}, z_{l,j}),$$

subject to

$$\sum_{l=1}^{k} u_{i,l} = 1, \quad 1 \leq i \leq n,$$

where U is an $n \times k$ partition matrix, $u_{i,l}$ is a binary variable with $u_{i,l} = 1$ indicating that object i belongs to cluster l. $Z = \{Z_1, Z_2, \ldots, Z_k\}$ is a set of k vectors representing the centroids of the k clusters. $d(x_{i,j}, z_{l,j})$ is a distance measure between object i and the centroid of cluster l on the jth variable. For numerical variables, $d(x_{i,j}, z_{l,j}) = (x_{i,j} - z_{l,j})^2$, and for categorical variables, $d(x_{i,j}, z_{l,j}) = 0$ if $x_{i,j} = z_{l,j}$ and 1 otherwise. If all variables are categorical, the algorithm is called k-modes. If both numerical and categorical variables are included in the data, the algorithm is called

k-prototypes. The optimization is achieved by solving the following two minimization problems iteratively:

1. Fix $Z = \hat{Z}$, and solve the reduced problem $P(U, \hat{Z})$,
2. Fix $U = \hat{U}$, and solve the reduced problem $P(\hat{U}, Z)$.

The first reduced problem is solved by

$$u_{i,l} = 1 \text{ if } \sum_{j=1}^{m} d(x_{i,j}, z_{l,j}) \leq \sum_{j=1}^{m} d(x_{i,j}, z_{t,j}) \text{ for } 1 \leq t \leq k;$$
$$u_{i,t} = 0 \text{ for } t \neq l.$$

The second reduced problem is solved by

$$z_{l,j} = \frac{\sum_{i=1}^{n} u_{i,l} x_{i,j}}{\sum_{i=1}^{n} u_{i,l}} \text{ for } 1 \leq l \leq k \text{ and } 1 \leq j \leq m, \text{ if variables are numerical, or}$$

$z_{l,j} = a_j^r$, if variables are categorical

where a_j^r is the mode of the variable values in cluster l.

In summary, the k-means algorithm partitions n objects into k clusters, where each of the objects is allocated to the cluster whose mean is nearest. Each time an object is reassigned, the centroids of both its old and new clusters are updated. The algorithm continues until no more reassignments take place.

Note that the final assignment of objects depends on the initial clustering seeds to a certain extent. Although most major reassignments will usually occur in the first several iterations, it is advisable to repeat the process, starting with different cluster seeds, to check the stability of the clusters.

A modified version of the k-means clustering is called the k-medoids algorithm, or sometimes "partitioning around medoids". It is robust to outliers, as it selects an object (medoid), rather than the centroid, to represent each cluster. This algorithm is implemented by carrying out the following two tasks:

1. Selecting objects as representatives (medoids) of clusters. A binary variable y_i is defined as $y_i = 1$ if and only if X_i is selected as a representative, and 0 otherwise.
2. Assigning X_j to one of the selected representatives. A binary variable $z_{i,j}$ is defined as $z_{i,j} = 1$ if and only if object X_j is assigned to the cluster whose representative is X_i, and 0 otherwise.

The clustering solution is optimized by minimizing the total dissimilarity, which is defined by the following function (Vinod, 1969):

$$\sum_{i=1}^{n} \sum_{j=1}^{n} d(i,j) z_{i,j},$$

subject to

$$\sum_{i=1}^{n} z_{i,j} = 1, \tag{3.5}$$

$$z_{i,j} \leq y_i,$$ (3.6)

and

$$\sum_{i=1}^{n} y_i = k, \quad \text{where} \quad i,j = 1, \ldots, n.$$ (3.7)

Constraint (3.5) indicates that each object must be assigned to only one cluster representative. Constraint (3.6) ensures that X_j can be assigned to X_i only if X_i has been selected as a cluster representative. The third constraint implies that k objects are selected as representatives of k clusters. Since the clusters are constructed by assigning each object to the closest representative, k nonempty clusters are formed in the final solution. The process of minimization consists of two phases. In the first phase an initial clustering is determined by selecting k representatives successively, while in the second phase the initial clustering is improved by swapping pairs of objects to reduce the total dissimilarity. One of the major advantages of this algorithm is that it considers all possible swaps, and therefore it is independent on the order of the objects to be clustered. The reader is referred to Kaufman and Rousseeuw (1990) for further discussions.

The use of the k-means and k-medoids algorithms is demonstrated here, using the simulated data in the previous section. The produced groups of those 3000 values are displayed in the figure below. It can be seen that both clustering procedures were able to produce groups that agree with the design of simulations. Furthermore, the grouping accuracy of the two methods is 92.83% and 92.54%, respectively, which are marginally higher than those of the average linkage and Ward's methods.

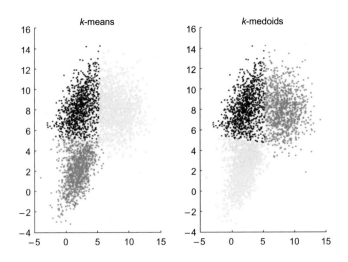

3.3.3   Deciding on the number of clusters

A major issue in cluster analysis, either hierarchical or nonhierarchical, is to decide on the number of clusters. Although there are no hard and fast rules, both qualitative and quantitative guidelines are available. For one thing, theoretical, conceptual or practical considerations may suggest a certain number of clusters. For instance, if the purpose of clustering is to identify market segments, managers may be interested in a particular number of clusters. For another, numerous quantitative procedures have been proposed in the literature to facilitate decisions on the number of clusters. For the details of such procedures, the reader is referred to the survey conducted by Milligan and Cooper (1985).

A widely applied quantitative procedure, named the Silhouette coefficient, is described here. Proposed by Rousseeuw (1987), it takes into account both cohesion and separation of clusters. The computation of the Silhouette coefficient can be outlined in the following three steps:

1. For the ith object, calculate its average inter-point distance to all the other objects in its cluster, denoted D_{ic_i}, where c_i denotes the cluster that contains the ith object.
2. For the ith object, calculate its average inter-point distance to all objects in a cluster which does not contain this object, denoted D_{ij} where $j \notin c_i$.
3. The Silhouette coefficient of the ith object is computed as

$$s_i = \frac{\min_j\{D_{ij}\} - D_{ic_i}}{\max(\min_j\{D_{ij}\}, D_{ic_i})}, \quad j \notin c_i.$$

For a determined clustering solution, the overall "goodness of fit" is evaluated by the average Silhouette coefficient:

$$SC = \frac{1}{n}\sum_{i=1}^{n} s_i,$$

where n is the number of objects. The range of s_i is $[-1, 1]$. Positive Silhouette coefficients are desirable, implying that the within-cluster variation is of a smaller magnitude than the between-cluster variation. The average within-cluster distance is desired to be as close to zero as possible, as in such cases the produced clusters are considerably dense.

In practice, for those datasets without a known number of clusters, it is advisable to repeat the clustering process over different values of k, and compute the SC scores accordingly. The k value corresponding to the highest SC score is considered as the optimal number of clusters. Note that k must take values between 2 and $n-1$, as the SC is not defined when $k = 1$, and it is always equal to one when $k = n$.

The *SC* score is useful to evaluate the effectiveness of a clustering solution. Kaufman and Rousseeuw (1990) suggested the following interpretation:

- If $SC \leq 0.25$, there is no substantial clustering structure;
- If $0.25 < SC \leq 0.50$, the clustering structure is weak and could be artificial;
- If $0.50 < SC \leq 0.70$, there is a reasonable clustering structure;
- If $SC > 0.70$, a strong clustering structure has been found.

Again, using the simulated data in the previous sections, the computation of Silhouette coefficients is demonstrated. For the sake of simplicity, only the *k*-means algorithm is implemented. 2-, 3-, 4- and 5-cluster structures are considered, producing four clustering solutions. The Silhouette coefficients are computed for individuals, which are plotted in the figure below. It is observed that most of these coefficients are positive, but none of them are fairly close to 1. This implies that the cohesion of the produced clusters is not very good.

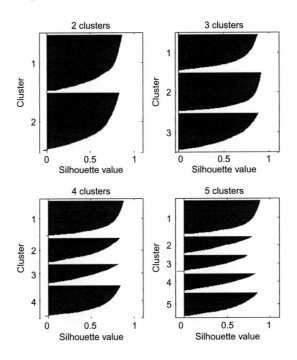

The computed average Silhouette coefficients are reported as follows:

	2 clusters	3 clusters	4 clusters	5 clusters
SC score	0.645	0.692	0.614	0.578

Based on these values, it is concluded that a 3-cluster structure is the best, which agrees with the simulation design. Since the 3-cluster *SC* score is very close to 0.7, the produced clusters are fairly reliable.

3.4 Fuzzy clustering

The clustering techniques discussed above are applicable when the underlying data generating process (model) is fixed. In such cases, each observation belongs to one group only. These grouping procedures are labelled as "crisp clustering", as they produce crisp groups, that is, the defined groups are mutually exclusive. However, as claimed by D'Urso and Maharaj (2009), data often display dynamic features, which imply the possibility of "group switching". This is especially the case when the group boundaries are not clear-cut. For example, consumers who do not have a preference for one soft drink brand over another could belong to either the Coke group or the Pepsi group, and thus they tend to switch groups very often. In this case, crisp clustering procedures are unlikely to reveal the underlying structure of data.

In practice, such "group switching" patterns should be taken into account when carrying out data clustering. This motivates the proposal of fuzzy clustering approaches, which do not assign observations exclusively to only one group. Instead, an individual is allowed to belong to more than one group, with an estimated degree of membership associated with each group. The degrees of membership take values between zero and one, indicating the degree to which an individual belongs to various groups. For example, a consumer could be assigned to the Coke, Pepsi and Sprite groups at the same time, with membership degrees 0.4, 0.35 and 0.25, respectively.

Numerous fuzzy clustering approaches have been proposed in the literature. See D'Urso and Maharaj (2009) for details. One method that was not mentioned in their study, but has been widely employed, is the Gaussian mixture models (GMM). Banfield and Raftery (1993) defined the term "model-based cluster analysis" for clustering based on finite mixtures of Gaussian distributions and related methods. Following the notations of Coretto and Hennig (2010), the GMM is briefly introduced here. For simplicity, univariate GMMs are considered, and the generalization to multivariate cases is straightforward.

Assume that real-valued observations $x_1, \ldots, x_n$ are modelled as independently identically distributed (i.i.d.) with the density

$$f(x; \theta) = \sum_{j=1}^{G} \pi_j \phi(x_i; \mu_j, \sigma_j^2),$$

where G denotes the number of mixture components, $\phi(\cdot, \mu_j, \sigma_j^2)$ is the density of the jth Gaussian distribution with mean μ_j and variance σ_j^2, and π_j is the proportion of the jth mixture component satisfying $\sum_{j=1}^{G} \pi_j = 1$ and $\pi_j \geq 0$ for any j. If G is unknown, the Bayesian information criterion can be employed as a standard estimation method. θ denotes the parameter vector consisting of G, $\mu_1, \ldots, \mu_G$ and $\sigma_1, \ldots, \sigma_G$, which are estimated by the maximum likelihood estimator:

$$\hat{\theta}_n = \text{argmax}_{\theta \in \Theta} \sum_{i=1}^{n} \log f(x_i; \theta),$$

where $\Theta = \left\{ \theta | \sigma_j^2 \geq s, \ j = 1, \ldots, G, \ \sum_{j=1}^{G} \pi_j = 1 \right\}$ for some choice of $s > 0$ that represents the lower bound of variances to avoid degeneracy of the log-likelihood function. The maximum likelihood estimation is usually carried out using the expectation–maximization (EM) algorithm. The EM algorithm follows an iterative process involving two steps, namely, the Expectation step (or E-step) and the Maximization step (or M-step). In the E-step, a function is created to describe the expectation of the log-likelihood evaluated using the current parameter estimates; in the M-step, updated parameter estimates are computed by maximizing the log-likelihood expectation function that was defined in the previous E-step. These new parameter estimates are then employed in the next E-step. The iteration continues until convergence.

Once $\hat{\theta}_n$ is obtained, the posterior probability that the ith observation x_i was generated by the lth component is computed by

$$\hat{\tau}_{i,l} = \frac{\hat{\pi}_l \phi(x_i; \hat{\mu}_l, \hat{\sigma}_l^2)}{f(x_i; \hat{\theta})},$$

and for n observations to be clustered into G groups, an n-by-G posterior matrix can be constructed:

$$\hat{\tau} = \begin{pmatrix} \hat{\tau}_{1,1} & \cdots & \hat{\tau}_{1,G} \\ \vdots & \ddots & \vdots \\ \hat{\tau}_{n,1} & \cdots & \hat{\tau}_{n,G} \end{pmatrix}.$$

It is important to note that the Gaussian mixture models are special cases of finite mixture models, which involve a family of probability density functions:

$$f(x; \theta) = \sum_{j=1}^{G} \pi_j p_j(x | \theta_j),$$

where $x = (x_1, \ldots, x_n)^T \in R^n$, and $p_j(\cdot)$ is a density function of the jth component parameterized by θ_j. The posterior probability that the ith observation was generated by the lth component is of the following form:

$$\hat{\tau}_{i,l} = \frac{\hat{\pi}_l p_l(x_i | \hat{\theta}_l)}{f(x_i; \hat{\theta})},$$

and the n-by-G posterior matrix can be constructed.

The posterior matrix $\hat{\tau}$ is used to assign observations to groups. Since $\hat{\tau}_{i,l}$ is the estimated probability of the ith observation being generated by the lth component, it is then considered as the likelihood of x_i belonging to Group l, as Group l is characterized by the lth component of the mixture models. If a crisp clustering procedure were used, the groups of $x = (x_1, \ldots, x_n)^T$ would have been determined as follows:

$arg \ \max_l \hat{\tau}_{i,l}$, where $l = 1, \ldots, G$.

That is, x_i is assigned to the group that produced the greatest posterior probability. In the fuzzy clustering setting, however, $\hat{\tau}_{i,l}$'s indicate the membership degrees of each observation, that is, to what extent an individual is believed to belong to each of the groups. This implies that each of the x_i's is allowed to be assigned to more than one group, where the fuzziness in the assignment procedure is evaluated by the $\hat{\tau}_{i,l}$'s. Especially for those individuals close to group boundaries, it is more plausible to consider their groupings in terms of membership degrees rather than in terms of total membership versus non-membership. Therefore, fuzzy procedures exhibit greater adaptivity in defining representatives of clusters, and greater sensitivity in capturing dynamic patterns that characterize data.

To illustrate fuzzy clustering using the Gaussian mixture models, the same simulated dataset, as used in the previous sections, is revisited. A three-component GMM is fitted, and the parameters are estimated by the EM algorithm. Convergence was achieved after 34 iterations, with log-likelihood $= -14334$. The parameter estimates are very close to the hypothetical values:

$$\hat{\mu}_1 = \begin{bmatrix} 1.93 \\ 1.97 \end{bmatrix}, \quad \hat{\Sigma}_1 = \begin{bmatrix} 1.92 & 1.42 \\ 1.42 & 2.82 \end{bmatrix};$$

$$\hat{\mu}_2 = \begin{bmatrix} 2.08 \\ 8.06 \end{bmatrix}, \quad \hat{\Sigma}_2 = \begin{bmatrix} 3.02 & 1.96 \\ 1.96 & 3.98 \end{bmatrix};$$

$$\hat{\mu}_3 = \begin{bmatrix} 8.08 \\ 7.90 \end{bmatrix}, \quad \hat{\Sigma}_3 = \begin{bmatrix} 3.71 & -0.05 \\ -0.05 & 3.11 \end{bmatrix}.$$

The estimated probability density of the three-component mixture distribution is visualized by the figure below. Note that the modes of the three components are distinct, but the "tails" overlap to some extent. As a result, for those values on the "tails", a fuzzy clustering method is appropriate.

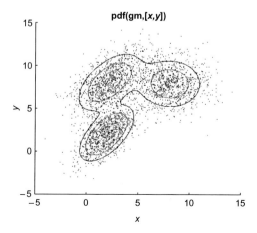

Knowing the parameter estimates, a 3000-by-3 posterior matrix can be constructed:

$$
\hat{\boldsymbol{\tau}} = \begin{pmatrix} \hat{\tau}_{1,1} & \hat{\tau}_{1,2} & \hat{\tau}_{1,3} \\ \vdots & \vdots & \vdots \\ \hat{\tau}_{3000,1} & \hat{\tau}_{3000,2} & \hat{\tau}_{3000,3} \end{pmatrix},
$$

and each row in this matrix consists of three membership degrees corresponding to the three groups, respectively. The following figure displays the posterior probabilities of Group 1, 2 and 3, that is, the first, second and third columns of $\hat{\boldsymbol{\tau}}$, respectively:

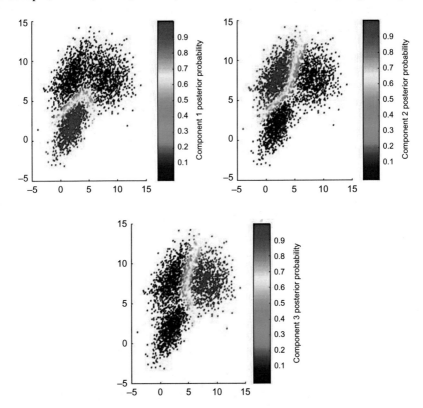

It can be seen that most points are either red or blue, that is, they are either very likely or very unlikely to belong to a certain group. Apparently, the grouping of these points is not questionable. On the other hand, the grouping of the other points is not clear-cut, as they are associated to more than one group.

Following the RGB colour model, the membership degree of each point is displayed, together with the estimated contour. It is obvious from the figure below that the points around the peaks are labelled by primary colours (red, green or blue), implying that these points have a fairly high membership degree associated to one of the three groups. In contrast, those points on "tail" areas are painted by mixed colours, indicating a mixture of memberships.

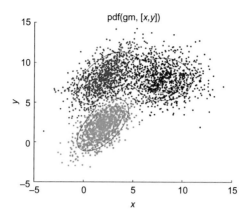

Now, a cut-off point for the membership degree needs to be specified to evaluate the crispness or fuzziness of the clusters. D'Urso and Maharaj (2009) suggested that for a three-group situation, one may consider 0.7 as the cut-off point. That is, if an individual has a membership value greater than 0.7, it is then classified into the corresponding group with a high level of confidence, as its combined membership in the other two groups is less than 0.3. Otherwise, there is a reasonable level of fuzziness associated with this individual, and its grouping remains unclear. In total, about 5% of the simulated data points exhibit fuzziness in their groupings. For the others, the clustering accuracy is 96.9%, which is considerably higher than those from the previous sections. It is worth noting that even when a crisp procedure is considered, the GMM still achieves desirable performance, resulting in 95.1% accuracy.

The fitted GMM can be used to classify newly observed data. 1000 additional data points are simulated from Group 1, and then their posterior probabilities are computed using the fitted GMM, which are displayed below. Based on the posterior probabilities, 952 of these data points were correctly classified into Group 1, 10 of them were misclassified into Group 2 and 5 of them were misclassified into Group 3. The remaining 33 observations exhibit fuzziness, as none of them has a membership greater than 0.7.

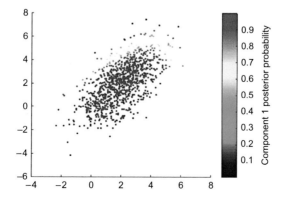

Appendix

R code for principal component analysis and factor analysis

```r
data <- read.csv("Evaluation.csv", header = T, sep = ",")
data <- data[,20:31]
cor.mat <- cor(data)
# Function for KMO
kmo <- function(x)
{
  x <- subset(x, complete.cases(x)) # Omit missing values
  r <- cor(x) # Correlation matrix
  r2 <- r^2 # Squared correlation coefficients
  i <- solve(r) # Inverse matrix of correlation matrix
  d <- diag(i) # Diagonal elements of inverse matrix
  p2 <- (-i/sqrt(outer(d, d)))^2 # Squared partial correlation coefficients
  diag(r2) <- diag(p2) <- 0 # Delete diagonal elements
  KMO <- sum(r2)/(sum(r2) + sum(p2))
  MSA <- colSums(r2)/(colSums(r2) + colSums(p2))
  return(list(KMO = KMO, MSA = MSA))
}

# Function for Bartlett test
Bartlett.sphericity.test <- function(x)
{
  method <- "Bartlett's test of sphericity"
  data.name <- deparse(substitute(x))
  x <- subset(x, complete.cases(x)) # Omit missing values
  n <- nrow(x)
  p <- ncol(x)
  chisq <- (1-n + (2*p + 5)/6)*log(det(cor(x)))
  df <- p*(p-1)/2
  p.value <- pchisq(chisq, df, lower.tail = FALSE)
  names(chisq) <- "X-squared"
  names(df) <- "df"
  return(structure(list(statistic = chisq, parameter = df, p.value = p.value,
      method = method, data.name = data.name), class = "htest"))

}

kmo(data)
Bartlett.sphericity.test(data)

# PCA
pca <- princomp(data, cor = TRUE) # Fit the model
summary(pca)
loadings(pca)
plot(pca,type = "lines") # Scree Plot
```

```
# FA
fit <- factanal(data, 2, scores = "Bartlett", rotation = "varimax") # Fit
the model with rotated factors, request factor scores
communalities <- (1 - fit$uniquenesses) # compute communalities
fit$loadings
fit$scores
load <- fit$loadings[,1:2]
rhat <- load%*%t(load) # Reproduced correlation matrix by 2 common factors
resid <- cor.mat - rhat # Difference between original and reproduced
version
round(rhat,2) # Display reproduced correlation matrix
round(resid,2) # Display the difference between original and reproduced
version
```

MATLAB code for cluster analysis

```
n = 1000; % No. of objects in each group
rng default % For reproducibility

% Group 1
mu1 = [2,2];
sigma1 = [2,1.5;1.5,3];
r1 = mvnrnd(mu1,sigma1,n);
% Group 2
mu2 = [2,8];
sigma2 = [3,2;2,4];
r2 = mvnrnd(mu2,sigma2,n);
% Group 3
mu3 = [8,8];
sigma3 = [4,0;0,3];
r3 = mvnrnd(mu3,sigma3,n);

figure;
plot(r1(:,1),r1(:,2),'.','Color','red')
hold on
plot(r2(:,1),r2(:,2),'.','Color','blue')
plot(r3(:,1),r3(:,2),'.','Color','green')
hold off

X = [r1;r2;r3];
T = [ones(n,1); 2*ones(n,1); 3*ones(n,1)]; % True clustering

dotsize = 3; % Dot size in plots

%% single, complete, average linkage and Ward's method
figure;
Z = linkage(X,'single'); % Single Linkage
c = cluster(Z,'maxclust',3);
subplot(221)
scatter(X(:,1),X(:,2),dotsize,c,'filled')
```

```matlab
title('Single Linkage')
[R_Single,~] = CLUSIND(T,3,c,3);

Z = linkage(X,'complete'); % Complete Linkage
c = cluster(Z,'maxclust',3);
subplot(222)
scatter(X(:,1),X(:,2),dotsize,c,'filled')
title('Complete Linkage')
[R_Complete,~] = CLUSIND(T,3,c,3);

Z = linkage(X,'average'); % Average Linkage
c = cluster(Z,'maxclust',3);
subplot(223)
scatter(X(:,1),X(:,2),dotsize,c,'filled')
title('Average Linkage')
[R_Average,~] = CLUSIND(T,3,c,3);

Z = linkage(X,'ward'); % Ward method
c = cluster(Z,'maxclust',3);
subplot(224)
scatter(X(:,1),X(:,2),dotsize,c,'filled')
title('Ward')
[R_Ward,~] = CLUSIND(T,3,c,3);

%% k-means and k-medoids
figure;
c = kmeans(X,3,'emptyaction','drop');
subplot(121)
scatter(X(:,1),X(:,2),dotsize,c,'filled')
title('k-means')
[R_kmeans,~] = CLUSIND(T,3,c,3);

c = kmedoids(X,3);
subplot(122)
scatter(X(:,1),X(:,2),dotsize,c,'filled')
title('k-medoids')
[R_kmedoids,~] = CLUSIND(T,3,c,3);

%% Compute the average Silhouette coefficients for k = 2, 3, 4 and 5
figure;
subplot(221)
c = kmeans(X,2,'emptyaction','drop');
[s2,h] = silhouette(X,c); S2 = mean(s2);
title('2 Clusters')

subplot(222)
c = kmeans(X,3,'emptyaction','drop');
[s3,h] = silhouette(X,c); S3 = mean(s3);
title('3 Clusters')

subplot(223)
c = kmeans(X,4,'emptyaction','drop');
```

```
[s4,h] = silhouette(X,c); S4 = mean(s4);
title('4 Clusters')

subplot(224)
c = kmeans(X,5,'emptyaction','drop');
[s5,h] = silhouette(X,c); S5 = mean(s5);
title('5 Clusters')

%% Gaussian mixture models
options = statset('Display','final');
gm = fitgmdist(X,3,'Options',options);

figure;
scatter(X(:,1),X(:,2),dotsize,'filled')
hold on
ezcontour(@(x,y)pdf(gm,[x y]),[-5,15],[-5,15]);
hold off

P = posterior(gm,X);
figure;
subplot(131)
scatter(X(c1,1),X(c1,2),dotsize,P(c1,2),'filled')
hold on
scatter(X(c2,1),X(c2,2),dotsize,P(c2,2),'filled')
scatter(X(c3,1),X(c3,2),dotsize,P(c3,2),'filled')
hold off
clrmap = jet(80); colormap(clrmap(9:72,:))
ylabel(colorbar,'Component 1 Posterior Probability')

subplot(132)
scatter(X(c1,1),X(c1,2),dotsize,P(c1,1),'filled')
hold on
scatter(X(c2,1),X(c2,2),dotsize,P(c2,1),'filled')
scatter(X(c3,1),X(c3,2),dotsize,P(c3,1),'filled')
hold off
clrmap = jet(80); colormap(clrmap(9:72,:))
ylabel(colorbar,'Component 2 Posterior Probability')

subplot(133)
scatter(X(c1,1),X(c1,2),dotsize,P(c1,3),'filled')
hold on
scatter(X(c2,1),X(c2,2),dotsize,P(c2,3),'filled')
scatter(X(c3,1),X(c3,2),dotsize,P(c3,3),'filled')
hold off
clrmap = jet(80); colormap(clrmap(9:72,:))
ylabel(colorbar,'Component 3 Posterior Probability')

figure;
scatter(X(:,1),X(:,2),6,P,'filled')
hold on
ezcontour(@(x,y)pdf(gm,[x y]),[-5,15],[-5,15]);
hold off
```

```
% Fuzzy clustering
cut = 0.7; % Cut-off value
I = max(P,[],2) < = cut;
sum(I)

TT = T;
TT(I) = [];
PP = P;
PP(I,:) = [];
[~,c] = max(PP,[],2);
[R_GMM_fuzzy,~] = CLUSIND(TT,3,c,3);

% Crisp clustering using GMM
c = cluster(gm,X);
c1 = c == 1;
c2 = c == 2;
c3 = c == 3;
figure;
scatter(X(c1,1),X(c1,2),dotsize,'r','filled');
hold on
scatter(X(c2,1),X(c2,2),dotsize,'b','filled');
scatter(X(c3,1),X(c3,2),dotsize,'k','filled');
hold off
[R_GMM_crisp,~] = CLUSIND(T,3,c,3);

% Classify new data using fitted GMM
Y = mvnrnd(mu1,sigma1,n);
PY = posterior(gm,Y);

figure;
scatter(Y(:,1),Y(:,2),dotsize,PY(:,2),'filled')
clrmap = jet(80); colormap(clrmap(9:72,:))
ylabel(colorbar,'Component 1 Posterior Probability')

I = max(PY,[],2) < = cut;
sum(I)
PP = PY;
PP(I,:) = [];
[~,c] = max(PP,[],2);
sum(c == 2) % No. of observations that were correctly classified into
Group 1
sum(c == 1) % No. of observations that were misclassified into Group 2
sum(c == 3) % No. of observations that were misclassified into Group 3
```

References

Banfield, J., & Raftery, A. (1993). Model-based Gaussian and non-Gaussian clustering. *Biometrics*, *49*, 803−821.

Bartlett, M. S. (1937). Properties of sufficiency and statistical tests. *Proceedings of the Royal Statistical Society, Series A*, *160*, 268−282.

Coretto, P., & Hennig, C. (2010). A simulation study to compare robust clustering methods based on mixtures. *Advances in Data Analysis and Classification, 4*, 111–135.

D'Urso, P., & Maharaj, E. A. (2009). Autocorrelation-based fuzzy clustering of time series. *Fuzzy Sets and Systems, 160*, 3565–3589.

Kaiser, H. (1970). A second generation little jiffy. *Psychometrika, 35*(4), 401–415.

Kaufman, L., & Rousseeuw, P. J. (1990). *Finding groups in data: An introduction to cluster analysis*. New York: John Wiley & Sons, Inc.

MacQueen, J. (1967). Some methods for classification and analysis of multivariate observations. *Proceedings of the Fifth Berkeley Symposium on Mathematical Statistics and Probability* (vol. 1, pp. 281–297). Berkeley, California: University of California Press.

Milligan, G. W., & Cooper, M. C. (1985). An examination of procedures for determining the number of clusters in a data set. *Psychometrika, 50*, 159–179.

Reddy, P. N., & Acharyulu, G. V. R. K. (2009). *Marketing research*. New Delhi: Excel Books.

Rousseeuw, P. J. (1987). Silhouettes: A graphical aid to the interpretation and validation of cluster analysis. *Computational and Applied Mathematics, 20*, 53–65.

Vinod, H. D. (1969). Integer programming and the theory of grouping. *Journal of the American Statistical Association, 64*(326), 506–519.

Ward, J. H., Jr. (1963). Hierarchical grouping to optimize an objective function. *Journal of the American Statistical Association, 48*, 236–244.

Computer vision in big data applications

4

About a half of our brain power is used for visual processing. This functionality has taken approximately 540 million years to evolve and develop. It used to be a consensus that only human beings and animals have the capability of processing visual images. However, machines in our contemporary society are also capable of acquiring, analyzing and understanding visual images. A well-programmed robot can easily recognize a bird in a picture and identify its colour, shape or species. This is known as computer vision, which is closely related to machine learning and artificial intelligence (AI).

Computer vision has countless applications. As stated by Fei-Fei Li, the leading AI scientist of Stanford University, robots with the ability to understand what they 'see' can serve human beings in a huge variety of areas, such as medical diagnosis, face recognition or verification system, video camera surveillance, transportation, etc. (Li et al., 2007). To achieve desirable performance, there are numerous AI scientists (not only academics like Fei-Fei Li but also those from Amazon, Facebook, Google or Microsoft) devoting massive resources to computer vision technology. Over the past decade, computer vision has been proven successful in factory automation (e.g. defective products detection, pick-and-place assistant) and medical image processing (e.g. automatic CT image segmentation). It should be stressed that all those applications must be carried out under controlled circumstances with planned routes, whereas technical challenges remain if a computer vision method is applied to a real-life environment where the illumination conditions may change or background clutters may exist.

Fortunately, because of the emergence of big data and complex machine learning algorithms in recent years, computer vision technologies are now capable of solving various real-life problems, such as vehicle registration plate identification. On a tollway, for instance, the registration plate of a vehicle is recognized and identified from the picture taken by the monitoring camera, and then the corresponding driver will be notified and billed automatically. Another example is that all registration plates of vehicles in Google Street View are automatically detected and blurred, for the sake of confidentiality. With the development of computer science, statistics and engineering, a growing number of advanced, exciting computer vision applications are approaching, such as driverless cars (e.g. Google self-driving cars), autonomous drones and household robots. It is worth noting that the success of these applications is rooted in the era of big data. In this chapter, we discuss how big data facilitate the development of computer vision technologies, and how these technologies can be implemented in big data applications.

Computational and Statistical Methods for Analysing Big Data with Applications.

4.1 Big datasets for computer vision

While high-performance computers and advanced machine learning algorithms may enhance the performance of image processing, computer vision can be facilitated by big datasets. In 2012, a large-scale image recognition contest was hosted by Stanford University (Deng, 2009). Despite the fact that the winning method can be programmed and trained to recognize various categories of objects in an image, the competition is made possible by a set of more than 14 million images collected by the Machine Vision Group of Stanford University (Deng, 2009). Such a huge dataset enables sufficient training of computer vision models, which consequently leads to greater accuracy of image recognition.

In the twenty-first century, big datasets of visual images are usually constructed by integrating resources from the internet. Images and videos are uploaded and shared through online platforms, while various datasets are being created online based on search engines and social media. Every day, the internet generates trillions of images and videos, while more than 85% of such content is multimedia imagery. The volume of photos and videos generated in the past 30 days is bigger than all images dating back to the dawn of civilization. In the presence of massive volume, large variety and high velocity, programming machines to recognize all objects in the observed images seems to be an impossible task to carry out. However, the performance of computer vision can be enhanced substantially by big data. Images obtained from various sources, such as websites, satellites, road cameras, remote sensors or mobile phones, can be considered as training datasets, which are the cornerstone of 'knowledgeable' computer vision algorithms that are capable of giving insights into all those images. As a consequence, the current community of computer vision is no longer interested in closed, in-house databases, whereas researchers are transferring their focus from small training samples to extensive datasets.

To construct a big dataset for image recognition, several key factors should be taken into account. Firstly, a big dataset should contain a large variety of object categories. The lack of diversity in the dataset is referred to as a fine-grained or subcategory image classification problem, which is not desirable in computer vision. The ultimate goal of computer vision is to endow machines/robots with the ability to observe and interpret the world, implying that it is essential to train those machines/robots properly so that they are capable of recognizing various types of objects under real-life circumstances. Since image processing is purely data driven, to some extent the history of computer vision can be summarized as the evolution of training datasets.

There are several milestones of multicategory image classification datasets that are worth reviewing. Early in 2004, Li, Fergus, and Perona (2007) created the Caltech 101 dataset which was among the first standardized datasets developed for multicategory image classification. The dataset was initially relatively small, consisting of 101 different object classes with 15−30 images per class. Those 101 categories range from small size objects (e.g. chairs) to large size items (e.g. aircrafts). It was later extended to 256 classes with background and scale variability. On the other hand, Deng (2009) created the so-called 'Big Data' ImageNet dataset (available at http://image-net.org/), which currently has 14,197,122 clean, annotated high-resolution images that are assigned to roughly 22,000 categories. As claimed by Russakovsky et al. (2014), the ImageNet dataset has larger scale and diversity than any other multicategory image classification dataset that was developed in the past decade. The ImageNet was created with the backbone of WordNet hierarchy, taking disambiguating word meanings and combines synonyms into the same class. The collection of ImageNet was conducted using both manual and automatic strategies on multiple online search engines.

The second key factor in the construction of a big dataset for image recognition is annotation. As a rule of thumb in the community of computer vision, annotations are important in training various types of models. Most annotations of ImageNet were implemented based on crowdsourcing technologies with both carefully designed annotation tools and automatic crowdsourced quality control feedback system (Su et al., 2012).

For model training purposes, most computer vision researchers consider only a subset of the ImageNet containing only 1000 categories, where each category has roughly 1000 images. Overall, there are about 1.2 million training images with 50,000 validation images and 150,000 for testing. This subset of images is employed by the ImageNet Large-Scale Visual Recognition Challenge (ILSVRC, which will be discussed in Section 4.6). The ImageNet is the biggest classification dataset so far. Any two of the selected 1000 classes neither overlap nor belong to the same parent node in the WordNet hierarchy, implying that subcategories are not considered. For example, various cat species are treated as one single class in the dataset. Each category is selected under the following eight constraints: object scale, number of instances, image clutter, deformability, amount of texture, colour distinctiveness, shape distinctiveness and real-world size.

The annotation of datasets enhances the flexibility of image processing. Generally speaking, there are two approaches to training an image classification model. The first approach, referred to as supervised training, assumes that the labels of images are known during the training process. The second approach, referred to as unsupervised training, assumes that the labels of images are unknown (labels are still needed for testing purposes). In order to support both supervised and unsupervised training schemes, the ImageNet dataset is annotated by adding a tight bounding box around the object of interest during training and testing processes so that the noise in the background of the image is minimized. This bounding box annotation is helpful when training an image detection model.

Two criteria are widely considered to evaluate the quality of bounding boxes. First, the tightness of bounding boxes would greatly affect machine learning algorithms when carrying out object recognition tasks. Normally better alignment turns into good performance. Second, in each image every object instance should be labelled. This would facilitate the algorithm to achieve a good coverage of objects with different instances.

As mentioned before, all annotations are done through a crowdsourcing platform. The whole annotation system is built on a large scale online 'human computing' platform named the Amazon Mechanical Turk (AMT). Similar to AMT, there are other well-known platforms such as Label Me and Annotatelt. These platforms let users upload their tasks for other annotators to label datasets in exchange for monetary credit. For ImageNet, the labels in the dataset are annotated by presenting the user with a set of candidate images, and then the user will provide binary answers as whether the image contains the candidate categories.

While there are numerous contributors on AMT, a quality control system is necessary to ensure a high quality of annotations. In practice, participants of AMT could make mistakes, or they may not follow the instructions strictly and therefore the labels for the same object may be inconsistent. To overcome such limitations, the quality control system is designed to take multiple users' inputs and make a final labelling decision only if most of those labels are consistent with each other. In order to pass a predefined confidence score, a minimum of 10 users are asked to perform labelling on each image. If the confidence level is not reached, more users are required to contribute to the labelling of this image.

Since occlusion occurs quite often, bounding box labelling is a more complicated task than simply giving labels to images. The system encourages users to place bounding around the object of interest, regardless of whether cluttering or occlusion exists. Three workers are necessary to provide a reasonable bounding box annotation: the first worker draws one bounding box around each object of interest on the given image; the second worker visually verifies the quality of bounding boxes; the third worker determines whether each object has a bounding box. Although such a three-step process is reasonably effective, the produced results cannot be guaranteed error-free. In fact, errors are nearly inevitable under particular circumstances, for example, when an object is too small in the image or a bounding box is too blurry.

To verify the annotations of the ImageNet, 10 object categories are randomly selected. In each category, the system randomly chooses 200 images to check whether those objects are correctly covered by bounding boxes. Results show that around 97.9% of those images are correctly annotated.

4.2 Machine learning in computer vision

To process massive visual information in big datasets like ImageNet, we need to employ methods that carry out image processing automatically. In computer science, machine learning techniques are designed for such purposes. Machine learning can be viewed as a set of methods that can automatically recognize patterns in data, while the recognized data patterns are used to predict newly observed values.

Machine learning has been a popular research field in the past several years. The main objective is to endow machines with the intelligence of human beings. It was evolved from the broad field of artificial intelligence. The outcome from decades of research is a large number of accurate and efficient algorithms that are convenient to use for even a practitioner. The core advantage of machine learning is that it is capable of providing predictions based on a huge amount of data, which is nearly an impossible task for a human being to perform. The long list of examples where machine learning techniques are extensively and successfully implemented includes text classification and categorization (spam filtering), network intrusion detection (cancer tissue, gene finding), brain computer interfacing, monitoring electric appliances, optimization of hard disk caching strategies and disk spin-down prediction, drug discovery, high-energy physics particle classification, recognition of hand writing and natural scene analysis. The idea of machine learning is to predict future states of a system before they are explicitly explored. In this context, learning is treated as an inductive interference with training samples representing incomplete information about statistical phenomena. Therefore, machine learning is not only a question of remembering but also the generalization to unseen cases.

As discussed above, a machine learning algorithm can be either unsupervised or supervised. Unsupervised learning typically aims to reveal hidden regularities or to detect anomalies in data, which is not commonly applied in computer vision. In image processing, supervised learning, sometimes referred to as classification, is of much interest. The aim of supervised learning is to predict the label of any novel observation based on the labelling rules that are learnt from training data.

A typical supervised machine learning algorithm consists of the following three steps:

- Data collection and annotation
- Feature Engineering
- Classification.

In the context of computer vision, we already learnt how visual image data are collected and annotated. In the next two sections, feature engineering and classification methods will be discussed respectively.

4.2.1 Feature engineering

Before implementing a computer vision algorithm, feature engineering is used to define the input of the algorithm manually, denoted X. When dealing with object recognition problems, instead of simply feeding raw pixels into a trained classifier, researchers tend to extract features from the raw image. An intuitive example is that objects like 'apples' or 'oranges' can be distinguished by particular features such as colour, size and shape. This process normally leads to better performance if one has better understanding of the problem, that is, objects are described by more appropriate features. Once suitable features are defined, one can design a system to quantify those features and then feed them into a classifier.

Over the past decade, computer vision scientists have designed various feature engineering systems to tackle image recognition problems. Before the emergence of automatic feature learning, computer vision techniques used to rely on feature descriptors that were designed manually, which are often time consuming. Moreover, manual feature engineering is usually subjective, implying that the user's experience or skills may influence the feature engineering process severely. Consequently, the performance of these engineered features tends to be questionable.

To enhance the reliability of manual feature engineering, attempts have been made over the past 20 years, resulting in a number of handcrafted feature benchmarks. Lowe (1999) introduced scale invariant feature transformation (SIFT), which has been the most famous feature engineering technique over the past two decades. The breakthrough it made was that it is a descriptor of local features, providing a robust solution to image patch matching with strong rotation tolerance. SIFT is computationally efficient, since it applies key point searching algorithms to reduce computational cost. In particular, at each key point that is predefined and detected by the difference of Gaussian method, SIFT generates a 128-dimensional feature vector. The local feature representation uses image gradient information, while the scale/rotation invariant property is achieved by binning and normalizing the principal direction and magnitude of local gradients. SIFT has much impact on geometry related problems such as stereo vision, motion recognition, etc. In fact, it has been evidenced to be a reliable system for wide-baseline stereo. Moreover, SIFT is the basis of the well-known Bags of Visual Words method (Csurka, 2004; Perronnin et al., 2010), which is very powerful in object recognition. However, if the aim is to identify matches between two distinct object instances from the same visual object category, SIFT cannot achieve desirable performance.

The Bags of Visual Words method is inspired by large-scale text matching. The main contribution of this method is that it interprets every local image patch as a visual word. Based on those words, it creates a codebook. Clustering methods such as k-means are used to learn the codebook of different words (centroids of k-means) in an unsupervised manner. Local SIFT features are employed to represent each word (image patch) and then a low dimensional real value vector is generated. Finally, each image can be represented by a histogram of those learned visual words. Numerous variants of this model and related computer vision implementations have been developed, for example, Spatial Pyramid Bags of Visual Words (Lazebnik, 2006) which introduces a hierarchical method for the integration of spatial information into feature representations.

Since the year 2005, an object template matching method, named Histogram of Oriented Gradients (HOG) (Dalal, 2005), has been receiving a growing amount of interest in the field of computer vision. HOG counts occurrences of gradient orientation in localized portions of an image, and it was initially designed for pedestrian detection. The popularity of HOG is because of its simple structure as well as fast processing speed. Its extended version, known as the Deformable Part Model (DPM) proposed by Felzenszwalb (2008), has also been widely applied. DPM is a system designed on top of the HOG features, which is an object detection technology based on a set of multiscale models. It consists of a coarse root filter that approximately covers the entire object and higher resolution filters that cover smaller parts of the object, where each filter is defined in a HOG feature space. In 2010, the DPM methodology claimed champion in the PASCAL Visual Object Classes challenge (http://host.robots.ox.ac.uk/pascal/VOC/), which is one of the top-class computer vision contests.

Since the time it was proposed, the HOG method had been dominating localized feature engineering until the emergence of deep convolutional neural networks

(CNNs), also known as deep learning. Deep learning (Donahue et al., 2012; Krizhevsky, Sutskever, & Hinton, 2012) has been playing a major role as a means of automatic feature learning, and it is referred to as the state-of-the-art methodology. The community of computer vision has witnessed breakthrough results of deep learning on various computer vision tasks such as object recognition, object detection, segmentation, etc. An extensive discussion of deep learning will be provided in Section 4.3.

4.2.2 Classifiers

Now we have reached the last step of machine learning, namely, classification. As discussed in Chapter 2 of this book, the aim of classification is to assign an object to one of several known groups of objects. In the context of computer vision, classification refers to the process that an object in a newly observed picture is identified as one of many known categories, which were learnt from training datasets. This highlights the importance of considering training datasets that contain a large variety of objects; if there is a lack of variety, the machine learning algorithm would not be able to recognize certain categories of objects. For example, if a training set of animal pictures only contains dogs, cats and birds, any new object would be classified only as a dog, a cat or a bird, even if it was a horse.

Classification techniques for computer vision are usually referred to as classifiers. In this section, we discuss three widely applied classifiers for image processing: regression, support vector machine (SVM) and Gaussian mixture models (GMM).

Regression

We start with a linear function, which is the simplest regression model. The linear regression equation can be written as follows:

$$y = f(x) = w^T x + \epsilon,$$

where x is the input matrix consisting of a number of dependent variables (covariates), w denotes the corresponding weights of x variables and ϵ is the residual term that represents the difference between predictions and real values of the response variable y. It is usually assumed that ϵ follows a normal distribution with zero mean and finite constant variance σ^2, i.e. $\epsilon \sim N(0, \sigma^2)$. The weights of x variables are determined properly so that the 'overall' difference between the predicted and real values of y is minimized.

For example, suppose we want to construct a linear regression model where the response variable is the price of a vehicle. To determine w, a sample of vehicle prices is collected, as well as particular features of those vehicles such as make, mileage, horsepower, etc. Obviously, y is the collection of vehicle prices while x consists of the observed features. The predicted prices, denoted $\hat{y}$, is then computed as

$$\hat{y} = \hat{w}^T x,$$

and the difference between y and $\hat{y}$ is expected to be as small as possible. Optimal values of w are determined by minimizing the following cost function:

$$J(w) = \frac{1}{n}\sum (y - w^T x)^2.$$

In other words, given a training dataset, the regression model with optimal w values would result in the lowest rate of training error.

The linear regression model is suitable for predicting continuous variables like vehicle prices. To work with discrete variables (e.g. binary classification results such as 0 and 1), the model can be generalized to cope with a binary classification setting. This is achieved by applying the sigmoid function, which is also known as a squashing function since real values are mapped onto [0, 1]. This mapping can be expressed by the following equation:

$$\text{sig}(w^T x) = \frac{1}{1 + \exp(-w^T x)},$$

where the weights w are optimized so that if y belongs to Class 1, the probability $P(y = 1|x) = f(x)$ is as large as possible while the probability $P(y = 0|x) = 1 - f(x)$ is as small as possible. The cost function in this case is defined as follows:

$$J(w) = -\sum (y \log(f(x)) + (1 - y)\log(1 - f(x))).$$

The binary classification model stated above can be further generalized to deal with multicategory classification problems. Assume that the labels of classes are now denoted $\{1, \ldots, K\}$ where K is the number of classes. A multinomial logistic regression (sometimes known as the softmax regression) is then employed to estimate the probability of y belonging to each class. Consequently, the output of a multinomial logistic regression is a vector consisting of K probability values that are associated with the K classes, respectively:

$$f(x) = \begin{bmatrix} P(y = 1|x, w) \\ P(y = 2|x, w) \\ \vdots \\ P(y = K|x, w) \end{bmatrix},$$

where w denotes the weights of covariates. Similarly as before, w is optimized so that the rate of training error reaches the lowest level, or equivalently, the cost function $J(w)$ is minimized.

Note that in a K-category classification exercise $(K \geq 2)$, the cost function $J(w)$ is a non-convex function. To minimize this function in an efficient way, we briefly discuss a method named stochastic gradient descent. Given a set of starting values of parameters, this method aims to update parameter values at each iteration so that the cost function value is getting closer to a local optimum. The iteration continues until a predefined convergence rule is satisfied. Mathematically, the stochastic gradient descent method can be described as follows:

$$w = w - \alpha \nabla E[J(w)],$$

where α is the learning rate parameter which is typically smaller than 0.1. The merit of this method is that it can be trained in mini batches, implying that the weights can be updated with a relatively small amount of training data in each iteration. This is advantageous to those who work with large scale datasets where the entire training set cannot fit into the memory. Generally, the stochastic gradient descent method can result in good convergence to a local optimum. Nonetheless, it cannot guarantee a global optimum after the convergence condition is satisfied. Moreover, it is worth noting that if the input data are somehow ordered, one should shuffle the dataset for batch training; otherwise the computed gradient at each training loop might be biased, leading to poor convergence eventually.

For regression-based feature learning examples, the interested reader is referred to the following website: https://github.com/amaas/stanford_dl_ex.

Support vector machine

SVM is an extremely popular machine learning algorithm for image processing. SVM conceptually implements the following idea: generate a decision boundary with large margins, which is the smallest distance between sample points to the separation boundary. Over the past decade, large margin classification techniques have been mainstreamed gradually, since large margins tend to label all samples correctly.

SVM is facilitated by kernel techniques, as kernel functions can be employed to solve optimization problems in high-dimensional spaces. When implementing SVM, training data are mapped onto a new feature space (normally with a higher dimension) using a kernel function, and then SVM produces a large margin separation between training samples in the new feature space. Given labelled sequences $(x_1, y_1), \ldots, (x_m, y_m)$ where x denotes the covariates and $y \in \{-1, 1\}$ is the response, a kernel function k is utilized by SVM in the following equation:

$$f(x) = \sum_{i=1}^{m} a_i k(x_i, x) + b,$$

where the coefficients a_i and b are estimated by minimizing the following function:

$$\sum_{i,j=1}^{m} a_i a_j k(x_i, x_j) + C \sum_{i=1}^{m} \zeta_i,$$

subject to

$$y_i f(x_i) \geq 1 - \zeta_i,$$

where ζ_i measures the degree of misclassification of x_i, and C is the penalty parameter of misclassification. The function $f(x)$ maps the training vectors x onto a higher, or even infinite dimensional space. Based on $f(x)$, SVM determines a linear hyper plane that separates training samples with maximized margin in the higher dimensional space. The reader is referred to Cortes and Vapnik (1995) for further details.

Gaussian mixture models

In Section 3.4 of this book, we discussed GMM as a fuzzy clustering tool. In the field of computer vision, GMM is widely applied as a means of soft classification, which is conceptually similar to fuzzy clustering. For example, when implementing the Bags of Visual Words method, GMM is employed to assign cropped image patches to predetermined groups in a fuzzy manner. The fuzziness of the assignment is determined by occupation probabilities, which evaluate how likely one observation belongs to one particular class of objects.

Suppose an image I is cropped into K patches. The overall probability of Image I is of the following form:

$$P(I|\mu, \sigma) = \prod_{k=1}^{K} \sum_{c=1}^{C} w_c \phi(o_k; \mu_c, \sigma_c^2),$$

where o_k represents the kth cropped patch of Image I, ϕ denotes the probability density function of a normal distribution, w_c, μ_c and σ_c denote the weight, mean and standard deviation of the cth Gaussian component, respectively. The estimation of model parameters was discussed in Section 3.4. The occupation probability of the kth cropped patch belonging to the cth Gaussian component is computed as

$$\frac{w_c \phi(o_k; \mu_c, \sigma_c^2)}{\sum_{c=1}^{C} w_c \phi(o_k; \mu_c, \sigma_c^2)}.$$

When classifying images, the distribution of local inputs from images is modelled by GMM. For each object class, a Gaussian model is estimated using a set of training images that belong to that class. The highest computed occupation probability indicates the label of image being classified.

4.3   State-of-the-art methodology: deep learning

I think AI (artificial intelligence) is akin to building a rocket ship. You need a huge engine and a lot of fuel. If you have a large engine and a tiny amount of fuel, you won't make it to orbit. If you have a tiny engine and a ton of fuel, you can't even lift off. To build a rocket you need a huge engine and a lot of fuel. The analogy to deep learning (one of the key processes in creating artificial intelligence) is that the rocket engine is the deep learning models and the fuel is the huge amounts of data we can feed to these algorithms.

Andrew Ng

As a means of pattern recognition, deep learning has been proven very powerful (Donahue et al., 2012; Krizhevsky et al., 2012). Intuitively, deep learning can be viewed as a set of algorithms that can 'mimic the brain'. A brain has billions of neurons that are connected to each other, whereas a deep learning algorithm works with multiple layers of neurons that are connected in similar ways. The philosophy of deep learning is that it builds a hierarchy of complex concepts on the basis of simpler ones.

The great potential of deep learning was demonstrated by Google in 2012, where the possibility that machines are able to teach themselves recognizing certain items was evidenced. Researchers designed an experimental deep learning system consisting of 16,000 CPUs in parallel, which was fed by 10 million random, unlabelled images extracted from YouTube videos. At the end of the experiment, the deep learning system was able to identify whether an object is a cat or not, even without human training.

Compared to a brain, Google's neural system is relatively simple. Instead of a very complex connection structure, deep learning algorithms consider the connection among neurons in adjacent layers only. That is, neurons in the first layer may be connected with those in the second layer, but not in the third or fourth. In addition, signals being communicated among neurons travel in only one direction, that is, messages may be conveyed from the first layer to the second, but not in the opposite way. Such a network of neurons is called a 'feed-forward back-propagation network'.

There are two factors that allow machines to solve intuitive problems. First, machines are able to understand the world in terms of a hierarchy of concepts, that is, a sophisticated concept is defined in terms of relatively simple concepts, which are further defined in terms of simpler concepts. Second, the hierarchy of concepts is built 'bottom up'. For example, a young child recognizes a dog by its shape, colour and texture, which are subcomponents of the dog.

Having learnt its mechanism and capability, researchers now believe that deep learning is suitable for situations that involve a large amount of data and complex relationships between parameters. In this section, we discuss deep learning algorithms and demonstrate how they can be applied to solve large scale image recognition problems. Since a deep learning algorithm can be described as a complex network of neurons, we first introduce a single-neuron model, followed by a discussion of multilayer neural networks. The training process of a neural network is discussed at the end of this section.

4.3.1 A single-neuron model

Denote a training sample (x, y), where x is the input vector and y denotes the corresponding labels (also known as supervisory signals). A neural network defines a complex, non-linear form of hypotheses $h_{W,b}(x)$ with weights W and bias b, which need to be learnt from training data.

The simplest form of a neural network is a single-neuron model. The following diagram visualizes a single-neuron model with three inputs:

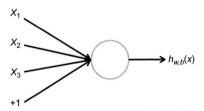

The neuron, denoted by the circle in the plot, is a computational unit that takes inputs x_1, x_2 and x_3, as well as a ' $+1$ ' term known as the bias. The output is computed as $h_{W,b}(x) = f(W^T x) = f\left(\sum_i^3 W_i x_i + b\right)$, where f is an activation function which may take various forms. Widely applied activation functions include the sigmoid function (as discussed before):

$$f(z) = \text{sig}(z) = \frac{1}{1 + \exp(-z)},$$

and the hyperbolic tangent function:

$$f(z) = \tan h(z) = \frac{\exp^z - \exp^{-z}}{\exp^z + \exp^{-z}},$$

which is a rescaled version of the sigmoid function with a wider range $[-1, 1]$, instead of $[0, 1]$.

4.3.2 A multilayer neural network

A complex neural network is constructed by linking multiple neurons so that the output of a neuron can be the input of another. Neurons are arranged in different layers, forming a multilayer neural network. The following figure displays an example of multilayer neural networks:

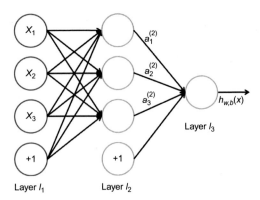

The bias term ' + 1' appears in both layers, corresponding to the intercept term. The leftmost layer of the network is called the input layer, while the rightmost layer is referred to as the output layer. Neurons in the middle layer are termed hidden units, as their values are not observed during the training process. This neural network has two sets of parameters, namely, $(W, b) = (W^1, b^1, W^2, b^2)$. Denote W_{ij}^l the parameter that is associated with the connection between unit j in layer l and unit i in layer $l + 1$, and denote a_i^l the activation of unit i in layer l. The neural network hypothesis $h_{W,b}(x)$ is then expressed as

$$h_{W,b}(x) = f\left(W_{11}^2 a_1^2 + W_{12}^2 a_2^2 + W_{13}^2 a_3^2 + b_1^2\right),$$

which can be either a real number or a score. This model explains the fundamentals of neural networks, which can be generalized to much more sophisticated structures. By constructing a network with more than one hidden layer, it is able to project inputs into non-linear combinations of features.

4.3.3 Training process of multilayer neural networks

A neural network can be trained using the stochastic gradient descent method, which greatly reduces the computational cost of the training process. To understand how the network is trained and how it differs from training a linear regression model, we discuss feed-forward pass and back-propagation pass.

Feed-forward pass

The cost function related to feed-forward pass is of the following form:

$$J = \frac{1}{2} \sum_{n=1}^{N} \sum_{k=1}^{K} (t_k^n - y_k^n)^2,$$

where N refers to the total number of training images, K represents the total number of object classes, y_k^n and t_k^n are vectors where t_k^n denotes the kth dimension of the nth pattern's corresponding label and y_k^n is the true label. Since the 'overall' cost of the entire training dataset is equal to the summation of all individual errors, the back-propagation can be carried out with respect to a single training sample at a time. Say for the nth training sample, the cost function is:

$$J^n = \frac{1}{2} \sum_{k=1}^{K} (t_k^n - y_k^n)^2.$$

To propagate an error to early layers, we need to take derivatives of the cost function with respect to the weights. To do so, the output of the current layer is computed as

$$u^{\mathscr{L}} = W^{\mathscr{L}} x^{\mathscr{L}-1} + b^{\mathscr{L}},$$

where $\mathscr{L}$ denotes the current layer, $x^{\mathscr{L}-1}$ denotes the input from the previous layer, $W^{\mathscr{L}}$ is the weights, and $b^{\mathscr{L}}$ denotes the bias. To map continuous values onto [0, 1], the sigmoid function is applied to produce the final output of layer $\mathscr{L}$:

$$x^{\mathscr{L}} = \text{sig}\left(u^{\mathscr{L}}\right).$$

Back-propagation pass

The most common learning algorithm for artificial neural networks is back propagation (BP), which refers to backward propagation of errors. To implement a BP neural network algorithm, input values are fed to the first layer and the signals are allowed to propagate through the network. To train a BP network, one must provide a learning set that consists of input samples and the corresponding known, correct output values. That is, the input-output pairs 'teach' the network what is correct and what is not. Each connection from one neuron to another has a unique weight, which is adjustable in the BP network because of the backward propagation of errors. If the network does not recognize a particular pattern correctly, the error is propagated backward and the weights are adjusted accordingly. The 'intelligence' of the network builds on the optimized weights, which result in the lowest rate of learning error. The ultimate goal of training is to determine a set of weights that enables the network to recognize patterns with 100% accuracy.

The 'errors' or 'costs' being propagated backward are related to the 'sensitivities' of each operating neural unit and perturbations of bias. This can be formulated as

$$\delta = \frac{\partial J}{\partial u}\frac{\partial u}{\partial b}.$$

To ensure that δ is back-propagated from higher layers to lower layers, we have the following equation:

$$\delta^{\mathscr{L}-1} = \left(w^{\mathscr{L}}\right)\delta^{\mathscr{L}} \cdot \text{sig}'\left(u^{\mathscr{L}}\right),$$

where $\delta^{\mathscr{L}}$ is the derivative of the layer $\mathscr{L}$ with respect to the operating neural units, and $\cdot$ denotes element-wise multiplication.

Now recall the stochastic gradient descent method that was applied to regression models. For a multilayer neural network, updating a weight assigned to a neuron depends on both the current and previous layers. The change in a neural weight is expressed as:

$$\Delta w = -\alpha x^{\mathscr{L}-1}\left(\delta^{\mathscr{L}}\right)^{T},$$

where $(\delta^{\mathscr{L}})^{T}$ denotes the transpose of $\delta^{\mathscr{L}}$.

So far we have learnt how a fully connected multilayer neural network operates. In the next section, we discuss CNNs which are central to computer vision applications.

4.4 Convolutional neural networks

'This is LeCun's hour', said Gary Bradski, an artificial intelligence researcher who was the founder of Open CV, a widely used machine vision library of software tools. Convolutional neural networks have recently begun to have impact because of the sharply falling cost of computing, he said, 'In the past there were a lot of things people didn't do because no one realized there would be so much inexpensive computing power available'.

Wired Magazine

Convolutional neural networks (CNN, or ConvNet) were initially proposed by LeCun et al. (1989) to recognize handwritten notes. A CNN is a type of feed-forward artificial neural network with individual neurons tiled in such a way that they respond to overlapping regions in the visual field. CNNs carry out subsampling of images so that computing time can be reduced. At each convolutional layer, feature maps from the previous layer are convolved with learnable filters, which then go through a transaction function to form new feature maps. Each newly generated feature map can be viewed as a combination of multiple input maps.

To understand how a CNN operates, we first discuss several simple concepts of a convolutional layer, as well as the motivation behind them. If we feed an image into an input layer, it is fully connected to all hidden units in the next layer. If the size of image is not too big (say, 20×20 pixels), it is not difficult to train the network since the number of parameters is relatively small. However, for larger images we need a neural network that may contain millions of parameters to describe them. For example, if a training dataset consists of 100 images, each of which has a 256×256 pixel resolution (this is not even a high resolution in reality!), then the number of parameters is approximately 6.5 million ($256 \times 256 \times 100$), leading to extremely high computational cost during the training process. One possible solution to this problem is to consider locally connected networks, where each neuron in hidden layers is connected to a subset of input units only. In image processing applications, this is implemented by connecting each hidden neuron to a small contiguous region of pixels in the input image, which can be explained as a filtering process (like convolutional filtering) in signal processing.

Due to natural properties of images, descriptive statistics of one part of an image tend to be similar to those of other parts. This implies that certain statistical features (e.g. kernels) learnt from a segment of image are applicable to other segments. When processing a large image, kernels learned from small patches of this image that are sampled randomly can be convolved, obtaining a different feature activation at each location in the image. Multiple feature maps can be produced for the image. In general, we have

$$x_j^{\mathscr{L}} = \text{sig}\left(\sum_{i \in M_j} x_j^{\mathscr{L}-1} k_{i,j}^{\mathscr{L}} + b_j^{\mathscr{L}}\right),$$

where M_j represents a subset of selected feature maps, $k_{i,j}^{\mathscr{L}}$ is the kernel convolved with inputs, and b denotes the bias.

4.4.1 Pooling

To reduce further the number of features produced by convolutional layers, an additional subsampling layer is needed to work with convolved feature maps. This is known as pooling. Pooling methods such as max-pooling or average-pooling are very useful when generating statistical features over a small region. Recall that images tend to have the 'stationarity' property, that is, features that are applicable in one subregion are likely to be applicable in other subregions. Pooling of features is carried out in such a way that, for example, the convolved features are divided into $M \times N$ disjoint regions and then the average of feature activation over these regions is computed. This process significantly reduces the dimension of features.

The following equation explains how pooling is performed:

$$x_j^{\mathscr{L}} = \mathrm{sig}(\beta_j^{\mathscr{L}} pool\left(x_j^{\mathscr{L}-1}\right) + b_j^{\mathscr{L}}),$$

where the term *pool* represents a pooling function. Usually this pooling function operates over all subregions with $M \times M$ size, implying that the output feature map is M times smaller than the original along both dimensions. Each output map has a unique bias term.

4.4.2 Training a CNN

Recall that when training a fully connected multilayer neural network we considered feed-forward and back-propagation passes. These techniques are applicable when training a CNN. To carry out the training, one needs to compute the gradients being propagated backward. As introduced in the previous section, a typical convolution framework has a convolutional layer followed by a down-sampling (pooling) layer. These layers are treated separately as Layer $\mathscr{L}$ and Layer $\mathscr{L} + 1$.

In order to compute the derivative of units at the current layer $\mathscr{L}$, one needs to sum over derivatives in the next layer with respect to units that are connected to the node in $\mathscr{L}$ that is of interest. An alternative way to interpret this process is that each unit in the feature map of $\mathscr{L}$ only corresponds to a block of pixels in the output map of $\mathscr{L} + 1$. Mathematically, the derivative being propagated backward is of the following form:

$$\delta_j^{\mathscr{L}} = \beta_j^{\mathscr{L}+1}(\mathrm{sig}'\left(u^{\mathscr{L}}\right) \cdot \delta_j^{\mathscr{L}+1}),$$

which is the sensitivity of a single feature map. The overall cost function is constructed as follows:

$$\frac{\partial J}{\partial b_j} = \sum_{u,v} \delta_j^{\mathscr{L}}{}_{u,v},$$

where u and v denote the vertical and horizontal positions in the feature map, respectively. The cost function with respect to kernels is obtained by summing the gradients for a given weight over all connections:

$$\frac{\partial J}{\partial k_{i,j}^{\mathcal{L}}} = \sum_{u,v} \delta_j^{\mathcal{L}}{}_{u,v} \left(patch_i^{\mathcal{L}-1}{}_{u,v} \right),$$

where $patch_i^{\mathcal{L}-1}{}_{u,v}$ represents the patch multiplied by the kernel $k_{i,j}^{\mathcal{L}}$ during convolution which forms a part of the feature map.

4.4.3   An example of CNN in image recognition

The research paper of Krizhevsky et al. (2012), entitled *ImageNet classification with deep convolutional neural networks*, presents a cutting-edge technology of computer vision. In this research, the authors trained a deep CNN consisting of eight layers (five convolutional layers and three fully connected layers) using the ImageNet dataset. On the basis of an optimized parallel training process, the CNN took approximately 1 week to learn the ImageNet, which was much faster than previous attempts. This methodology was the winner in the ImageNet Large Scale Visual Recognition Challenge 2012 (http://www.image-net.org/challenges/LSVRC/2012/), and it is now very popular in the community of computer vision. In this section, we briefly review this major breakthrough in the field of image recognition.

To tackle a problem of categorizing thousands of different objects from millions of images, Krizhevsky et al. (2012) proposed to use CNNs since CNNs have great learning capacity. In addition, CNNs can be controlled by varying the depth and width of each layer, exhibiting great flexibility in real-life applications. Moreover, since the ImageNet has a huge size, CNNs can facilitate the training process as they have much fewer parameters to be optimized, compared to fully-connected multi-layer neural networks.

Overall structure of the network

The CNN considered by Krizhevsky et al. (2012) consists of eight layers. The first five layers are convolutional layers while the last three are fully-connected layers. The prediction is generated by the softmax layer of the network, which is a process very similar to a multinomial logistic regression.

The first convolutional layer has 96 kernels of size $11 \times 11 \times 3$, where 11×11 denotes the resolution of a two-dimensional image and 3 denotes the three RGB channels of pixels. The stride of kernels is 4 pixels. The second convolutional layer consists of 256 kernels of size $5 \times 5 \times 48$, and therefore the number of feature maps is expected to be 256. A max-pooling layer is attached after the first and second convolutional layers. The third convolutional layer contains 384 kernels of size $3 \times 3 \times 256$, while the fourth convolutional layer has 384 kernels of size $3 \times 3 \times 192$. The last convolutional layer has 256 kernels of size $3 \times 3 \times 192$.

No pooling layer is connected to the third, fourth or fifth convolutional layer. It is observed that the size of kernels is getting smaller from the first convolutional layer to the last, since the size of output feature maps is decreasing. As a matter of fact, local information might be overlooked if large-size kernels are applied. As learnt before, the sigmoid or tanh function is usually employed as a standard transaction function. However, neither the sigmoid nor the tanh function is computationally efficient. As a consequence, the rectified linear unit (ReLU) is applied to the output of each convolutional and fully-connected layer. Mathematically, the ReLU takes the following form:

$$ReLU(x) = \max(0, x),$$

which is extremely useful when large-scale datasets (like ImageNet) are considered for training CNNs.

Data preprocessing

The input layer of a CNN has a fixed size, and consequently all input images need to be rescaled to fixed values of height and width. The side effect of the rescaling process is that some objects may lose their normal aspect-ratio, while shape information may be distorted. One way to alleviate this problem is to rescale the image so that the shorter side fits the required size. During the training process, each input image is down-sampled to a size of 256×256 with centre patch 224×224 cropped. The average value of the entire training dataset are calculated and subtracted from each input image. To boost the number of training images, each sample is mirrored through horizontal reflection. To augment the dataset further, Gaussian noises with zero mean and 0.1 standard deviation are randomly added to colour channels. The pre-added noises can make the system more robust to illumination and colour intensity variations.

Prevention of overfitting

When training a CNN, one important task is to avoid overfitting. In the context of image processing, overfitting refers to the situation that a trained model concentrates too much on image noises rather than general patterns in the image, which must be prevented to ensure reasonable robustness of pattern recognition. To do so, Dahl, Sainath, and Hinton (2013) proposed a method named 'dropout', which is applied to the output of two fully-connected layers in a CNN to prevent severe overfitting. The performance of the 'dropout' method can be evaluated by the dropout rate. For instance, a 0.5 dropout rate of one layer indicates that 50% of randomly selected neurons are shut down for both forward and backward passes, implying that these neurons will not contribute to the prediction of the CNN. This technique significantly reduces complex coadaptations of neurons by presenting a different architecture at input layers, which forces the CNN to learn more robust features instead of relying on features of certain neurons. In other words, the robustness of the CNN is enhanced by considering a smaller number of parameters during the training process.

We now have reviewed the basics of a CNN. In the next section, a tutorial is provided to demonstrate how CNNs are applied to solve real-life problems.

4.5 A tutorial: training a CNN by ImageNet

In this section, a tutorial is provided to demonstrate how a deep neural network, namely the CNN proposed by Krizhevsky et al. (2012) is trained. The required software package and computing code are discussed so that the interested readers are able to train their own CNNs.

4.5.1 Caffe

Training a CNN requires a deep learning framework named Caffe, which was developed by Jia et al. (2013) at the Berkeley Vision and Learning Center (http://caffe.berkeleyvision.org/). Under the BSD 2-Clause license (https://github.com/BVLC/caffe/blob/master/LICENSE), it is released and made available to public.

Caffe was created with expression, speed and modularity in mind. This modifiable deep learning framework is built on C++ library with Python and MATLB wrappers for training and deploying general purpose CNNs. It supports GPU (graphics processing unit)-accelerated computing and the implementation of Compute Unified Device Architecture (CUDA), being able to process up to 250 images per minute on a single Nvidia GPU. A comprehensive toolkit is available for training and testing exercises.

Detailed information about how Caffe can be installed on a Linux or OS X operating system is available at http://caffe.berkeleyvision.org/installation.html.

4.5.2 Architecture of the network

The Caffe deep learning framework is constructed of multiple layers, each of which is represented by a 4-dimensional array called 'blob'. A blob is capable of performing both a forward pass to produce inputs for the next layer and a backward pass that calculates the gradient with respect to the inputs and weights of the network. Caffe produces a complete set of layers including convolutional layers, pooling layers, fully-connected layers, local response normalization (LRN) layers, as well as loss functions.

The following table represents the CNN architecture that is of interest, where 'conv' denotes a convolutional layer and 'fc' stands for a fully-connected layer. For each convolutional layer, the number of convolutional filters is specified in the first row, while the second row indicates the convolutional stride and spatial padding. Pooling size is displayed in the third row, while max pooling is applied for the given down-sampling factor. The fourth row of a convolutional layer indicates whether LRN is applied or not. For a fully-connected layer, the first and fourth rows specify the size of receptive field and the dropout rate, respectively. Note that the layers 'fc6' and 'fc7' are regularized using a 50% rate of dropout, while the layer 'fc8' is a multiway softmax classifier. A ReLU is employed as the activation function for all layers except the final output layer, namely, 'fc8'.

conv1	conv2	conv3	conv4
$96 \times 11 \times 11$	$256 \times 5 \times 5$	$384 \times 3 \times 3$	$384 \times 3 \times 3$
(stride:4, padding:0)	(stride:2, padding:2)	(stride:1, padding:1)	(stride:1, padding:1)
Pooling $\times$ 3	Pooling $\times$ 3	—	—
LRN	LRN	—	—

conv5	fc6	fc7	fc8
$256 \times 3 \times 3$	4096	4096	500
(stride:1, padding:1)	—	—	—
Pooling $\times$ 3	—	—	—
—	Drop-out 0.5	Drop-out 0.5	Softmax

As Caffe is a machine learning framework that operates in an 'end-to-end' manner, a typical model within such a framework starts with an input layer, and ends with a layer that computes the objective function. The following paragraphs describe how various layers in the model are constructed, with relevant C++ code provided.

Input layer

The input layer of Caffe is constructed as follows:

```
layers {
  name: "data"
  type: DATA
  bottom: "data"
  top: "label"
  data_param {
    source: "data_train_lmdb"
    backend: LMDB
    batch_size: 350
  }
}
```

This layer is named as "data" because it provides the system with information about training data. The data parameter section "data_param {...}" carries out data loading tasks, reading training data into the system. The "source" command specifies the correct path to training data, while the "backend: LMDB" command implies that the training dataset (namely, ImageNet) is stored in the database named LMDB (lightning memory-mapped database). The "batch_size" specification depends on whether a central processing unit (CPU) or a GPU is in use. Moreover, the "bottom" and "top" commands specify the input and output of this layer, respectively.

Convolutional layer

```
layers {
  name: "conv1"
  type: CONVOLUTION
  bottom: "data"
  top: "conv1"
  blobs_lr: 1
  blobs_lr: 1
  convolution_param {
    num_output: 96
    kernel_size: 11
    stride: 4
  }
}
```

This layer is named "conv1", indicating that it is the first convolutional layer. Similarly as before, the "convolution_param {...}" section specifies particular parameter values. "num_output: 96" indicates that 96 kernels are considered with a size of 11 as specified by the "kernel_size: 11" command, and the magnitude of stride is set to 4. In addition, two learning rates need to be selected, corresponding to the weights of neurons and the bias term, respectively. Both rates are set equal to one as we want to update all parameters during back-propagation in the training process.

To normalize a convolutional layer, the ReLU is considered as a normalizing function, which can be programmed as follows:

```
layers {
  name: "relu1"
  type: RELU
  bottom: "conv1"
  top: "conv1"
}
```

Pooling layer

A pooling layer in Caffe can be formed as follows:

```
layers {
  name: "pool1"
  type: POOLING
  bottom: "conv1"
  top: "pool1"
  pooling_param {
```

```
      pool: MAX
      kernel_size: 3
      stride: 2
   }
}
```

The name "pool1" means it is the first pooling layer. The "pool: MAX" command refers to the max pooling method, with kernel size and stride parameters specified by the "kernel_size: 3" and "stride: 2" commands, respectively. The "bottom" and "top" lines indicate that "conv1" and "pool1" are the input and output of this layer, respectively.

LRN layer

A LRN layer is constructed on top of the max pooling layer "pool1":

```
layers {
   name: "norm1"
   type: LRN
   bottom: "pool1"
   top: "norm1"
   lrn_param {
      local_size: 5
      alpha: 0.0001
      beta: 0.75
   }
}
```

Note that "local_size", "alpha" and "beta" are known as normalization parameters. The 'pooling-normalization' process is repeated for each convolutional layer.

Fully-connected layers

A full connection between two layers is equivalent to the inner-product operation. Consequently, the first fully-connected layer is constructed as follows:

```
layers {
   name: "fc6"
   type: INNER_PRODUCT
   bottom: "pool5"
   top: "fc6"
   blobs_lr: 1
   blobs_lr: 1
   inner_product_param {
      num_output: 4096
   }
}
```

Note that the input of this layer is specified by the command "`bottom: "pool5""`", implying that this function connects the last convolutional layer with fully-connected layers. The ReLU of this layer is programmed as

```
layers {
  name: "relu6"
  type: RELU
  bottom: "fc6"
  top: "fc6"
}
```

Similarly, the second fully-connected layer is programmed as

```
layers {
  name: "fc7"
  type: INNER_PRODUCT
  bottom: "fc6"
  top: "fc7"
  blobs_lr: 1
  blobs_lr: 1
  inner_product_param {
    num_output: 4096
  }
}
layers {
  name: "relu7"
  type: RELU
  bottom: "fc7"
  top: "fc7"
}
```

Dropout layers

For the layers "fc6" and "fc7", dropout layers are required to prevent overfitting, which are constructed as follows:

```
layers {
  name: "drop6"
  type: DROPOUT
  bottom: "fc6"
  top: "fc6"
  dropout_param {
    dropout_ratio: 0.5
  }
}
layers {
  name: "drop7"
  type: DROPOUT
  bottom: "fc7"
```

```
top: "fc7"
dropout_param {
    dropout_ratio: 0.5
  }
}
```

where the rate of dropout is specified by the "dropout_ratio: 0.5" command.

Softmax layer

After "fc6" and "fc7", the last fully-connected layer "fc8" is constructed and connected to a softmax layer that gives the final classification score:

```
layers {
  name: "fc8"
  type: INNER_PRODUCT
  bottom: "fc7"
  top: "fc8"
  inner_product_param {
    num_output: 1000
  }
}
layers {
  name: "loss"
  type: SOFTMAX_LOSS
  bottom: "fc8"
  bottom: "label"
  top: "loss"
}
```

"num_output" in "fc8" is equal to 1000 because there are 1000 categories of objects in the ImageNet training dataset. In the 'softmax-loss' layer, labels of all training and validation images must be provided, as indicated by the "bottom: "label"" command.

Now we have all those eight layers in the CNN constructed. We then discuss the training process of the CNN.

4.5.3 Training

We first introduce a solver file, which determines whether the CNN is trained on a GPU or a CPU. With respect to parameter optimization, the maximum number of iterations is specified in the solver file as well. Typically, it has the following form:

```
net: "imagenet_arch.prototxt"
# Base learning rate, momentum and the weight decay of the network
base_lr: 0.001
momentum: 0.9
gamma: 0.1
```

```
weight_decay: 0.0005
# The maximum number of iterations
max_iter: 50000
# Solver mode: CPU or GPU
solver_mode: GPU
```

The first line "net: "imagenet_arch.prototxt"" points to the CNN architecture "imagenet_arch" that was created in the previous section. "base_lr", "momentum", "gamma" and "weight_decay" refer to the base learning rate, momentum, gamma and weight decay parameters that impact on the speed of learning and the convergence of the model. These values are determined following Krizhevsky et al. (2012). "max_iter: 50000" indicates that the optimization algorithm is run with 50,000 iterations at most, while the "solver_mode: GPU" command requires a GPU for processing the training. Note that the directory of this solver file is ".prototxt", which is the same as that of the constructed architecture.

Given those specified parameter values, Caffe trains the CNN using a stochastic gradient descent algorithm which is computational efficient. Training images are processed in a batch-by-batch manner, where the number of images being processed in each iteration depends on the size of batch. To start the training process, we simply type in a one-line command:

```
path_to_tool/caffe train --solver=solver
```

After every 100 iterations, the loss rate is displayed, which will eventually drop to a number less than 0.1. Once the CNN is trained, it can be utilized for prediction purposes. The interested reader is referred to http://caffe.berkeleyvision.org/tutorial/ for more examples.

4.6 Big data challenge: ILSVRC

The ILSVRC is considered as the most prominent contest in the field of computer vision. As mentioned in Section 4.1, it is based on a subset of the ImageNet dataset that consists of 1.2 million training images. In the ILSVRC, there are three tasks to perform: (i) object classification which aims to produce a list of objects in an image, (ii) single-object localization which concentrates on producing a bounding box around a particular object, and (iii) object detection which is a combination of (i) and (ii), that is, producing a list of categories of objects as well as the corresponding bounding boxes. In this section, we review briefly the winning methods in the history of ILSVRC.

4.6.1 Performance evaluation

To evaluate the performance of a participating method in the ILSVRC, the following rules apply:

- The organizer of ILSVRC provides contestants with a training sample (a subset of ImageNet as mentioned earlier), and specifies a test sample in which the labels of objects are unknown to contestants

- Each contestant uses the training sample for model optimization, and then the trained model is used to recognize unlabelled objects in the test sample
- For each unlabelled object, a contestant is allowed to make five guesses, that is, to identify five categories that have the greatest chance to contain the unlabelled object
- A contestant receives one point if the unlabelled object belongs to one of its top-five predictions
- The overall rating of a contestant is computed as the ratio of total score to the size of test sample.

4.6.2   Winners in the history of ILSVRC

ILSVRC was held for the first time in 2010. The winner of ILSVRC 2010 was Lin, Lv, Zhu, and Yang (2011), who achieved an error rate of 0.28, that is, 72% of the unlabelled objects were correctly categorized. Lin et al. (2011) used the SIFT and LBP features (Ojala, Pietikainen, & Maenpaa, 2002) with local coordinate coding and super vector coding representations as well as a multiclass stochastic SVM for classification, which has been a very popular method since then. The error rate of Lin et al. (2011) was reduced to 0.257 by the XRCE team (Xerox Research Centre Europe, http://www.xrce.xerox.com/Research-Development/Computer-Vision), who were the winners of ILSVRC 2011. Note that the categories of objects were not the same in ILSVRC 2010 and ILSVRC 2011, but it is not unreasonable to compare those two error rates as the number of categories and the size of training dataset remained the same.

The year 2012 had been a turning point of ILSVRC as deep neural networks emerged in the competition. Apart from the winning team SuperVision (Krizhevsky et al., 2012), the runner-up ISI team also achieved good performance by incorporating graphical Gaussian vectors (Harada & Kuniyoshi, 2012) into a linear classifier named passive-aggressive algorithm (Crammer, Dekel, Keshet, Shalev-Shwartz, & Singer, 2006).

Because of the success of deep learning in ILSVRC 2012, most participants of ILSVRC 2013 considered deep CNNs. The winner of this year was Clarifai, who employed an ensemble model that takes the average of predictions from multiple neural networks. The architecture of Clarifai is very similar to that of Krizhevsky et al. (2012), consisting of convolutional and fully-connected layers. Additional training data were employed to enhance the performance of classification, leading to an 11.2% error rate.

The number of ILSVRC participants continued increasing in 2014. Apart from research teams of universities, companies such as Google and Adobe also took part in the competition. While CNNs remained the dominating methodology, noticeable progress was made as the rate of classification error was significantly reduced. The winner of ILSVRC 2014 was GoogLeNet, a deep neural network proposed by a research team from Google (Szegedy et al., 2014). It consists of 22 layers, with multiple auxiliary classifiers connected to intermediate layers so that gradients can be backward-propagated efficiently through various layers. At the end of the competition, GoogLeNet resulted in a 6.7% classification error, which is smaller than that of any other method in the history of ILSVRC.

ILSVRC has witnessed major breakthroughs in computer vision since 2010, with greater accuracy of classification achieved by more sophisticated but efficient computing algorithms. So far, deep learning methodologies have been dominating the competition. It is hoped that the error of visual recognition will be reduced further in the forthcoming ILSVRC.

4.7 Concluding remarks: a comparison between human brains and computers

We conclude this chapter by highlighting the pros and cons of computer vision techniques. Russakovsky et al. (2014) carried out a comprehensive study of comparing human brains with machines that are capable of processing visual images. They noted that compared to machines, human beings are noticeably worse at recognizing subcategories of objects (e.g. Alaskan malamute and Husky, two dog breeds that look alike). This is probably because subcategories of objects are often distinguished by details (e.g. texture of birds) rather than overall features (e.g. shape of birds), whereas an 'untrained' human brain is not sensitive to insignificant changes in an image. Another reason may be that human beings do not usually experience a sufficient amount of images from subcategories. For example, if all the Labrador retrievers that a person has seen are either yellow or black, a chocolate one would not be recognized as a Labrador retriever by this person. Similar problems can be solved easily when training a machine, as long as the training sample has a large variety of objects.

On the other hand, however, machines tend to be outperformed by human brains under certain circumstances, which can be summarized as follows:

- Even the most powerful machine learning algorithm cannot recognize small objects in an image quite well. The reason is that computing algorithms base the learning process on the features of an object, which are extracted from pixels. If an object is represented by a small number of pixels, then the features that can be extracted would be insufficient and hence it would be rather difficult for a machine to recognize this object. In contrast, from a picture human beings are able to infer what an object would be, even if it is very small.
- Machine learning algorithms are sensitive to distortions in an image, such as contrast distortion, colour distortion or texture distortion. Such distortions lead to significant loss of information that can be utilized by a computing algorithm, whereas in many cases human beings are still able to tell what the object is. For instance, one would easily recognize that the following picture displays a distorted racing car, whereas a machine learning algorithm probably would not be able to identify this object.

- Machine learning algorithms are sensitive to changes in the background of an image. For example, if a CNN is trained by photos of dogs that were taken under standardized illumination conditions, it may not perform well when recognizing a dog from an image that has a brighter or darker background. This is apparently not a problem for human beings, as a change in illumination conditions hardly alters the initial decision.

- Last but not least, machines do not comprehend what they see. This fact is considered as the biggest gap between a robot and a human being. For example, a well-trained computing algorithm may be able to recognize the 'Mona Lisa' from other paintings of Leonardo da Vinci, but would not appreciate or critique it. However, with the assistance of big data, computers are becoming more and more knowledgeable, implying that the gap between robots and human beings is reducing.

Acknowledgements

We are grateful to Saizheng Zhang, Danfei Xu, Yang Li and Peng Ding for their comments and suggestions which have improved the quality and presentation of this chapter.

References

Cortes, C., & Vapnik, V. (1995). Support-vector networks. *Machine Learning, 20*(3), 273–297.

Crammer, K., Dekel, O., Keshet, J., Shalev-Shwartz, S., & Singer, Y. (2006). Online passive-aggressive algorithms. *Journal of Machine Learning Research, 7*, 551–585.

Csurka, G. C. (2004). Visual categorization with bags of keypoints. In *European conference on computer vision workshop*.

Dahl, G. E., Sainath, T. N., & Hinton, G. E. (2013). Improving deep neural networks for LVCSR using rectified linear units and dropout. In *International conference on acoustics, speech and signal processing*.

Dalal, N. A. (2005). Histograms of oriented gradients for human detection. In *Conference on computer vision and pattern recognition*.

Deng (2009). A large-scale hierarchical image database. In *Conference on computer vision and pattern recognition*.

Donahue, J., Jia, Y., Vinyals, O., Hoffman, J., Zhang, N., Tzeng, E., et al. (2012). Decaf: A deep convolutional activation feature for generic visual recognition. In *International conference on machine learning*.

Felzenszwalb, P. D. (2008). A discriminatively trained, multiscale, deformable part model. In *Conference on computer vision and pattern recognition*.

Harada, T., & Kuniyoshi, Y. (2012). Graphical Gaussian vector for image categorization. In *Conference on neural information processing systems*.

Jia, Y., Shelhamer, E., Donahue, J., Karayev, S., Long, J., Girshick, R., et al. (2013). *Caffe: Convolutional architecture for fast feature embedding*. Berkeley Vision and Learning Center.

Krizhevsky, A., Sutskever, I., & Hinton, G. E. (2012). Imagenet classification with deep convolutional neural networks. In *Conference on neural information processing systems*.

Lazebnik, S. C. (2006). Beyond bags of features: Spatial pyramid matching for recognizing natural scene categories. In *Conference on computer vision and pattern recognition*.

LeCun, Y., Boser, B., Denker, J. S., Henderson, D., Howard, R. E., Hubbard, W., et al. (1989). Backpropagation applied to handwritten zip code recognition. *Neural Computation*, *1*(4), 541−551.

Li, F. F., Fergus, R., & Perona, P. (2007). Learning generative visual models from few examples: An incremental bayesian approach tested on 101 object categories. In *Conference on computer vision and pattern recognition*.

Lin, Y., Lv, F., Zhu, S., & Yang, M. (2011). Large-scale image classification: Fast feature extraction and SVM training. In *Conference on computer vision and pattern recognition*.

Lowe, D. (1999). Object recognition from local scale-invariant features. In *Conference on computer vision and pattern recognition*.

Ojala, T., Pietikainen, M., & Maenpaa, T. (2002). Multiresolution gray-scale and rotation invariant texture classification with local binary patterns. *Pattern Analysis and Machine Intelligence*, *24*(7), 971−987.

Perronnin, F., Sanchez, J., & Mensink, T. (2010). Improving the fisher kernel for large-scale image classification. In *European conference on computer vision*.

Russakovsky, O., Deng, J., Su, H., Krause, J., Satheesh, S., Ma, S., et al. (2014). Imagenet large scale visual recognition challenge. *International Journal of Computer Vision*, *1*, 1−42.

Su, H., Jia, D., & Fei-Fei, L. (2012). Crowdsourcing annotations for visual object detection. In *Workshops at the twenty-sixth AAAI conference on artificial intelligence. 2012*.

Szegedy, C., Liu, W., Jia, Y., Sermanet, P., Reed, S., Anguelov, D., et al. (2014). Going deeper with convolutions. arXiv preprint:1409.4842.

A computational method for analysing large spatial datasets

Spatial datasets are very common in statistical analysis, since in our lives there is a broad range of phenomena that can be described by spatially distributed random variables (e.g., greenhouse gas emission, crop yield, underground water, per capita income, etc.). To describe these phenomena, one needs to collect information about the corresponding random variables at a set of locations within a particular region of study. The collected sample of data is a spatial dataset, which contains the coordinates of those locations and the values/features that were observed at these locations.

In practice, information about a random variable is collected at a limited number of locations. At those unsampled locations, a model of spatial data must be built so that the occurrence of similar outcomes can be predicted. To achieve this, conventional methods for the analysis of spatial data (e.g., sequential Gaussian simulation algorithms) have been applied frequently. However, these methods have several drawbacks. Of critical importance is the limitation that these methods usually require a large amount of random-access memory (RAM), leading to severe adverse impact on computational efficiency when analysing massive datasets. Moreover, these methods impose strong assumptions such as normality and linearity, treating spatial datasets as if they were straightforward to model. However, the spatial phenomena behind datasets usually exhibit considerable complexity, which goes beyond those imposed assumptions. To achieve good performance in data analysis, these assumptions need to be relaxed.

Liu, Anh, McGree, Kozan, and Wolff (2014) proposed a method using higher-order statistics (HOS) to characterize spatial random variables which are potentially sophisticated. In their proposal the HOS method was employed as a means of interpolation, while in general it can be considered as a tool for the exploration of a large variety of spatially distributed phenomena.

The HOS method is advantageous to analysts who work with large spatial datasets for the following reasons:

- It is computationally efficient. The HOS method reduces computational cost by introducing the concept "active replicates" of a spatial template. Unlike conventional approaches, the HOS method does not carry out loops over all possible pairs of data when characterizing spatial random fields. It considers only the most relevant, informative data points, disregarding those "not-so-relevant" ones. As a result, it does not require as much RAM as the other methods would do, and hence the computational efficiency is improved.
- It offers great flexibility. The HOS method is a distribution-free approach, implying that it can be applied to a broad range of datasets without imposing assumptions such as Gaussian

distribution or linear dependence. In addition, rather than a "point estimate + standard error" type of solution, the probability density function at any user-specified location can be approximated, providing full information about the prediction. Furthermore, it is applicable to both 2D and 3D cases.

- It is reliable, especially when the size of sample is large. The HOS method is completely data-driven, and it learns spatial patterns from the observed sample. The greater the sample size is, the more "knowledge" it can gain. Consequently, spatial patterns in a massive dataset can be represented without severe distortion.
- It is user-friendly. Firstly, it requires no training image, which implies that the user does not need to have prior knowledge about the spatial phenomenon being analysed. Secondly, the HOS method is robust to outliers, and hence the user does not need to undertake any preprocessing to deal with them.

The rest of this chapter is structured as follows: an introduction to spatial statistics is provided firstly, followed by a detailed discussion of the HOS method. Then the code of MATLAB functions that are used to implement this method are listed and discussed, and finally a case study is demonstrated briefly.

5.1   Introduction to spatial statistics

Let Z be a real-valued stationary and ergodic random field in $\mathbb{R}^d$. Let $Z_X(\cdot)$ be a random variable defined on Z, where the dot in the parentheses denotes spatial locations. The form of $Z_X(\cdot)$ implies that a change in its value can be explained by the change in its spatial coordinates (e.g., longitude and latitude). One of the most important tasks in spatial statistics is to model the dependence structure of spatially distributed variables. That is, one needs to determine an approximation or statistical emulation of the relationship between $Z_X(\cdot)$ values and the spatial locations at which those values were observed. This is known as spatial dependence modelling.

5.1.1   Spatial dependence

Tobler (1970) introduced the "First Law of Geography", which states "Everything is related to everything else, but near things are more related than distant things". In statistical terms, Tobler's Law refers to positive spatial autocorrelation in which pairs of observations that are taken nearby are more alike than those taken further apart, which implies that the core task of spatial dependence modelling is to describe the variation in $Z_X(\cdot)$ values as a function of the distance between two locations. To implement this, numerous techniques have been proposed in the literature. For illustration purposes we take the variogram as an example, which is the most widely applied method in spatial statistics. Cressie (1993) stated that under the assumption that Z is stationary, the variogram of $Z_X(\cdot)$ is defined as the variance of the difference between values at two locations across realizations of $Z_X(\cdot)$, which is of the following form:

$$2\gamma_X(s, v) = Var(Z_X(s) - Z_X(v)),$$

where $\gamma_X(s, v)$ is known as the semi-variogram. Define the distance $h = \|s - v\|$, the semi-variogram is written as:

$$\gamma_X(s, v) = \gamma_X(h).$$

If a sample of spatial data is available, the semi-variogram can be computed empirically. Denote $Z_X = (Z_X(x_1), \ldots, Z_X(x_n))$ as a sample of $Z_X(\cdot)$ observed at locations $x_1, \ldots, x_n$, respectively. It is assumed that these locations are mutually different. The empirical semi-variogram is then computed as the average squared difference between two values of $Z_X(\cdot)$ that are h-distance apart:

$$\hat{\gamma}_X(h) = \frac{1}{|N(h)|} \sum_{(i,j) \in N(h)} |Z_X(x_i) - Z_X(x_j)|^2,$$

where $N(h)$ denotes the collection of pairs of observations satisfying $\|x_i - x_j\| = h$, and $|N(h)|$ denotes the number of such pairs. By computing $\hat{\gamma}_X(h)$ values with a sequence of distances (e.g., 0 km, 0.1 km, 0.2 km, ...), a function can be established to represent the relationship between the variance of the difference in value and the corresponding distance. An example of typical (semi-) variograms is visualized in the figure below:

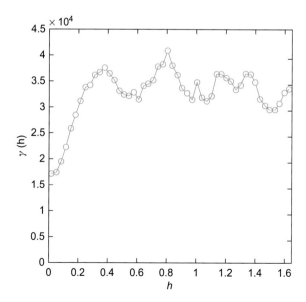

5.1.2 Cross-variable dependence

The variogram $\gamma_X(h)$ evaluates the spatial autocorrelation of $Z_X(\cdot)$ as a function of distance, that is, how the value of $Z_X(\cdot)$ at one location depends on the value of the same variable at another location as well as the distance between them. However, in many cases it is desirable to consider a secondary variable $Z_Y(\cdot)$, which exhibits cross-dependence with $Z_X(\cdot)$. That is, not only does the value of $Z_X(\cdot)$ at location s depend on its own value at location v, but it also depends on the values of $Z_Y(\cdot)$ at locations s and v. Taking into account the cross-variable dependence is often beneficial in analysing spatial datasets, as claimed by Goovaerts (1997). For example, in open-pit mining projects considering both copper and sulphur is more informative than considering copper only, since sulphur impacts on throughput at several processing stages; in environmental studies taking into account the information about land use or population density of a region has the potential to improve the prediction of water quality.

The cross-variable correlation between $Z_X(\cdot)$ and $Z_Y(\cdot)$ can be quantified by generalizing $\gamma_X(h)$ in a straightforward manner to the cross-variogram, which can be defined in the following two ways:

$$2\gamma_{X,Y}(s, v) = Cov(Z_X(s) - Z_X(v), Z_Y(s) - Z_Y(v)),$$
$$2\gamma_{X,Y}(s, v) = Var(Z_X(s) - Z_Y(v)).$$

Note that the first definition requires observing data of $Z_X(\cdot)$ and $Z_Y(\cdot)$ at the same locations whereas the second does not. Similarly as before, by defining $h = \|s - v\|$ the empirical cross-variogram can be computed at various values of h.

5.1.3 Limitations of conventional approaches to spatial analysis

The variogram, cross-variogram and those methods built on them (e.g., kriging/ co-kriging models) are conventional in the literature of spatial statistics. Although a large number of past studies have demonstrated the usefulness and desirable properties of these methods (Babak, 2014; Babak & Deutsch, 2009; Hwang, Clark, Rajagopalan, & Leavesley, 2012), their drawbacks are not negligible. As argued by Liu et al. (2014), most of these methods assume that it is sufficient to use the first- and second-order statistics to characterize random fields, whereas in practice spatially distributed phenomena often exhibit sophisticated features of spatial dependence which cannot be accounted for adequately if statistics up to the 2nd order are considered only. This motivates the proposal of methods that incorporate higher-order statistics (HOS), which are more appropriate when the random field Z exhibits complex patterns such as non-Gaussian distributions or nonlinear dependence.

5.2 The HOS method

The idea of considering higher-order statistics in the analysis of spatial data originates from the development of the "multiple-point (MP) simulation" algorithms. According to Zhang, Switzer, and Journel (2006), MP simulation aims at capturing local patterns of variability from a training image and anchoring them to the image or numerical model being built. As noted by Boucher (2009) as well as Mustapha and Dimitrakopoulos (2010b), multiple-point methods substantially address many of the limits of conventional methods. Being different from those methods based on the two-point variogram models, the MP simulation algorithm utilizes local proportions read from training images, which allows the reproduction of complex multiple-point patterns, such as undulating channels of underground water. Mustapha and Dimitrakopoulos (2010b) stated that the ability to deal with complexity by MP methods is related to the introduction of the training images or analogues of the phenomenon under study. These serve as the source of underlying patterns, and a simulated realization should reproduce their probability of occurrence. Various types of MP simulation algorithms have been proposed, including (but not limited to) the SNESIM algorithm of Strebelle (2002), the FILTERSIM algorithm of Zhang et al. (2006), the SIMPAT algorithm of Arpat and Caers (2007), and the extend version of SNESIM proposed by Boucher (2009).

There are two major limitations of the MP simulation algorithms. Firstly, in general these algorithms are RAM-demanding, especially when the patterns in data show a large variety. This drawback prevents the use of the MP simulation algorithms when dealing with massive spatial datasets. Secondly, Mustapha and Dimitrakopoulos (2010b) argued that these algorithms lack a general mathematical framework, that is, they are not well-defined spatial stochastic modelling frameworks. To overcome these limitations, the concept "high-order spatial cumulant" was proposed by Dimitrakopoulos, Mustapha, and Gloaguen (2010), and then simulation algorithms based on this concept were developed by Mustapha and Dimitrakopoulos (2010a, 2010b) and Mustapha and Dimitrakopoulos (2011). In these studies, the high-order spatial cumulant-based method showed superiority over the MP simulation algorithms. However, this method is confined to the cases where the interest is in a single variable, leading to potential loss of information about multivariate spatial dependence, that is, cross-variable dependence.

In this section, methodological details of the HOS method are given. We discuss firstly how the higher-order statistical properties of a random field can be measured in the presence of secondary variables, and then we introduce an approach to the approximation of probability density function at a user-defined location.

5.2.1 Cross-variable high-order statistics

This section strictly follows Section 2 of Liu et al. (2014), where the cross-variable higher-order spatial statistics were proposed to achieve high-quality characterization of sophisticated spatial dependence patterns. The cross-variable statistics enrich the information that can be extracted from data, and hence spatial features can be measured with greater accuracy.

We define a $(t_0, t_1, \ldots, t_r)$-order spatial moment as follows:

$$m_{t_0, t_1, \ldots, t_r} = E\big(Z^{t_0}(x_0) Z^{t_1}(x_1) \ldots Z^{t_r}(x_r)\big),$$

where x_0 stands for the reference location, and $x_1, \ldots, x_r$ are r distinct locations in the neighbourhood of x_0. It can be seen that $m_{t_0, t_1, \ldots, t_r}$ is capable of characterizing how the value of a single spatial random variable (e.g., $Z_X(\cdot)$ or $Z_Y(\cdot)$) at the reference location x_0 depends on the values nearby, taking into account higher-order properties. If both $Z_X(\cdot)$ and $Z_Y(\cdot)$ are present, we define the cross-variable higher-order spatial moment as follows:

$$m_{t_0, t_1, \ldots, t_r}^{Z_X} = E\Big(Z_X^{t_0}(x_0) Z_{j_1}^{t_1}(x_1) \ldots Z_{j_r}^{t_r}(x_r)\Big),$$

where the superscript of m corresponds to the primary variable $Z_X(\cdot)$, and $j_1, \ldots, j_r$ can be either X or Y. In other words, the value of the primary variable $Z_X(\cdot)$ at the reference location may depend on the values of $Z_X(\cdot)$ or $Z_Y(\cdot)$ or both, at nearby locations. For instance, suppose $r = 3$ with locations x_1, x_3 associated with $Z_X(\cdot)$ and location x_2 with $Z_Y(\cdot)$. The (2, 2, 1, 1)-order cross-variable spatial moment is:

$$m_{2,2,1,1}^{Z_X} = E\big(Z_X^2(x_0) Z_X^2(x_1) Z_Y(x_2) Z_X(x_3)\big).$$

It should be stressed that $m_{t_0, t_1, \ldots, t_r}^{Z_X}$ can be generalized to account for more than one secondary variable, for example, $Z_Y(\cdot)$ and $Z_W(\cdot)$.

The empirical version of $m_{t_0, t_1, \ldots, t_r}^{Z_X}$ requires the quantification of the spatial relationship between the reference location x_0 and locations $x_1, \ldots, x_r$ in its neighbourhood. To do so, Dimitrakopoulos et al. (2010) introduced the "spatial template" concept. Denote h_i the distance from x_0 to x_i, $i = 1, 2, \ldots, r$. Define unit vectors $\vec{d}_i$ as the directions from x_0 to x_i, which are usually characterized by the azimuth (the rotation clockwise from the north) and dip (the rotation vertically from the horizontal plane). h_i and $\vec{d}_i$ jointly quantify the location of x_i in relation to the reference location, that is, $x_i = x_0 + h_i \vec{d}_i$.

Denote $T_{h_1 \vec{d}_1, \ldots, h_r \vec{d}_r}$ the r-direction spatial template. A set of $r + 1$ spatial locations $\{x_{k_0}, x_{k_1}, \ldots, x_{k_r}\}$ is considered satisfying this template if they can be re-expressed by x_{k_0} and $h_i \vec{d}_i$. That is,

$$\{x_{k_0}, x_{k_1}, \ldots, x_{k_r}\} \in T_{h_1 \vec{d}_1, \ldots, h_r \vec{d}_r} \quad \text{if} \quad x_{k_i} = x_{k_0} + h_i \vec{d}_i \quad \text{for } i = 1, 2, \ldots, r.$$

The empirical cross-variable higher-order statistics is then computed as follows:

$$m_{t_0,t_1,\ldots,t_r}^{Z_X} = \frac{1}{N_{T_{h_1\vec{d}_1,\ldots,h_r\vec{d}_r}}} \sum_{k=1}^{N_{T_{h_1\vec{d}_1,\ldots,h_r\vec{d}_r}}} Z_X^{t_0}(x_k)Z_{j_1}^{t_1}\left(x_k + h_1\vec{d}_1\right)\ldots Z_{j_r}^{t_r}\left(x_k + h_r\vec{d}_r\right),$$

where $N_{T_{h_1\vec{d}_1,\ldots,h_r\vec{d}_r}}$ denotes the number of sets of locations that satisfy the template.

5.2.2 Searching process

As mentioned in the last section, the computation of $m_{t_0,t_1,\ldots,t_r}^{Z_X}$ requires an exhaustive search to discover all sets of locations that satisfy the template. A step-by-step demonstration is provided in this section to describe explicitly how the spatial template is defined and how the searching is conducted.

Suppose we have the following sample of data, in a 2D space:

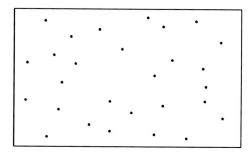

Consider an unsampled location as the reference, denoted x_0:

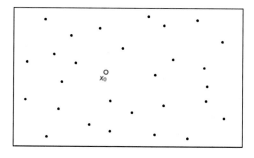

The observed data points are searched within the neighbourhood of x_0, and the corresponding locations are denoted x_1, x_2 and x_3, respectively:

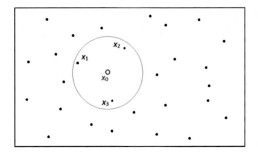

The number of observations in the neighbourhood is three, implying $r = 3$. Therefore, a 3-direction linkage from x_0 to x_1, x_2 and x_3 is considered as a template:

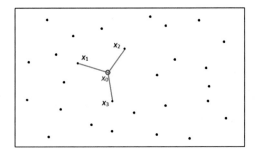

Since it is not reasonable to expect irregularly distributed data points falling exactly on the ends of the template, we should allow some tolerance for each template direction:

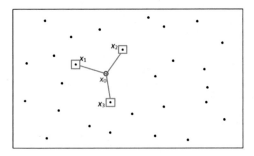

Now the spatial template at the unsampled location x_0, with respect to the three sampled locations in its neighbourhood, is determined:

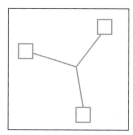

It is important to know that the determined spatial template remains unchanged for x_0 when carrying out the searching process. That is, as long as the searching is still with respect to x_0, the specifications of the template $(r, h_i$ and $\vec{d_i})$ should not be altered.

Once the template is determined, we carry out the exhaustive searching process by applying this template to each of the observed data points throughout the entire sample. To understand how this works, imagine that the template figure above is a transparent slide moving over the dataset:

Those data points that satisfy the template are considered as "active" replicates of the template, and will contribute to computing high-order spatial statistics. In the figure above, active replicates of the template cannot be observed at locations x_1 and x_4, as the surrounding data points do not satisfy the template. In contrast, we observe an active replicate at location x_5, since observations at x_6, x_7 and x_8 are within the tolerance of the predefined template.

Of course, to borrow a greater amount of information from the surrounding data, we can consider a bigger neighbourhood and hence a more sophisticated spatial template. For example, a 9-direction template at x_0 is as follows:

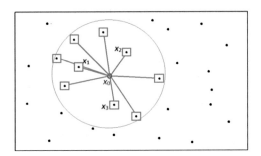

The next step is to record all active replicates of the template, which are used to compute the empirical cross-variable high-order spatial statistics, that is, to characterize the spatial relationship between the value at the unsampled location x_0 and those in its neighbourhood (namely, values at x_1, x_2 and x_3). Once this is completed, the probability density function at x_0 is approximated via polynomial expansion, which will be discussed in the next section.

5.2.3   Local CPDF approximation

The aim of this section is to approximate the probability density function at any user-specified location x_0, conditional on (i) the observed information in its surrounding region, and (ii) the characterized dependence features of the underlying random field. The former conditioning relies directly on the sample values of the primary and secondary variables (if available) in the neighbourhood of x_0, while the latter is on the basis of the computed cross-variable higher-order statistics.

Consider again Z, a real-valued stationary and ergodic random field in $\mathbb{R}^s$. In most practical situations $s = 2$ or 3. Suppose two samples of $Z_X(\cdot)$ and $Z_Y(\cdot)$ are observed respectively, denoted $\{Z_X(x_1), \ldots, Z_X(x_n)\}$ and $\{Z_Y(y_1), \ldots, Z_Y(y_m)\}$. That is, values of $Z_X(\cdot)$ are taken at locations $x_1, \ldots, x_n$ and those of $Z_Y(\cdot)$ are from $y_1, \ldots, y_m$. We assume that $x_i \neq x_j$ and $y_i \neq y_j$ for $i \neq j$, but we allow $x_i = y_j$ for some i and j, that is, it is possible to observe values of both random variables at one location.

We then introduce the HOS method, proposed by Liu et al. (2014), for the approximation of the conditional probability density function (CPDF) at any user-defined location, denoted x_0. The CPDF is of the following form:

$$f(Z_X(x_0)|\text{data}),$$

or equivalently

$$f(Z_X(x_0)|Z_X(x_1), \ldots, Z_X(x_n), Z_Y(y_1), \ldots, Z_Y(y_m)).$$

The expression above considers $Z_X(\cdot)$ as the primary variable and $Z_Y(\cdot)$ the secondary variable. Obviously, it could be the other way around:

$$f(Z_Y(y_0)|Z_X(x_1), \ldots, Z_X(x_n), Z_Y(y_1), \ldots, Z_Y(y_m)).$$

Without loss of generality, only the CPDF of $Z_X(x_0)$ is discussed. The CPDF can be expressed as follows, according to the Bayes rule:

$$f(Z_X(x_0)|Z_X(x_1), \ldots, Z_X(x_n), Z_Y(y_1), \ldots, Z_Y(y_m))$$
$$= \frac{f(Z_X(x_0), Z_X(x_1), \ldots, Z_X(x_n), Z_Y(y_1), \ldots, Z_Y(y_m))}{f(Z_X(x_1), \ldots, Z_X(x_n), Z_Y(y_1), \ldots, Z_Y(y_m))}.$$

Denote $f(\mathbf{Z}_X) = f(Z_X(x_0), Z_X(x_1), \ldots, Z_X(x_n), Z_Y(y_1), \ldots, Z_Y(y_m))$, that is, the numerator of the right hand side of the equation above. The approximation of the CPDF requires only the evaluation of $f(\mathbf{Z}_X)$, since $f(Z_X(x_1), \ldots, Z_X(x_n), Z_Y(y_1), \ldots, Z_Y(y_m))$ is obtained from $f(\mathbf{Z}_X)$ by integrating out $Z_X(x_0)$ over its support.

We approximate $f(\mathbf{Z}_X)$ by a finite-order series expansion of Legendre polynomials, as proposed by Mustapha and Dimitrakopoulos (2010a, 2010b). In brief, Legendre functions are solutions to the Legendre's differential equation. The pth-order Legendre polynomial is given by

$$P_p(x) = \frac{1}{2^p p!} \left(\frac{d}{dx} \right)^p \left[(x^2 - 1)^p \right] = \sum_{k=0}^{p} a_{k,p} x^k,$$

where $a_{k,p}$ is the coefficient of x^k given the order p, which is a constant. The detailed discussion of Legendre polynomials is out of the scope of this chapter. The interested reader is referred to Liu and Spiegel (1999). It is worth noting that the Legendre polynomials are orthogonal over the interval $[-1, 1]$.

Denote ω the finite order to which $f(\mathbf{Z}_X)$ is approximated. Then we have

$$f(\mathbf{Z}_X) \approx \sum_{i_0=0}^{\omega} \sum_{i_1=0}^{i_0} \cdots \sum_{i_n=0}^{i_{n-1}} \sum_{i_{n+1}=0}^{i_n} \cdots \sum_{i_{n+m}=0}^{i_{n+m-1}} L_{\bar{i}_0, \bar{i}_1, \ldots, \bar{i}_n, \ldots, i_{n+m}} \times \overline{P}_{\bar{i}_1}(Z_X(x_1)) \times \ldots \times \overline{P}_{\bar{i}_n}(Z_X(x_n))$$
$$\times \overline{P}_{\bar{i}_{n+1}}(Z_Y(y_1)) \times \ldots \times \overline{P}_{i_{n+m}}(Z_Y(y_m)) \times \overline{P}_{\bar{i}_0}(Z_X(x_0)) = \hat{f}(\mathbf{Z}_X),$$

where $\overline{P}$'s are normalized version of Legendre polynomials such that

$\overline{P}_p(x) = \sqrt{\frac{2p+1}{2}} P_p(x)$, $L_{\bar{i}_0, \bar{i}_1, \ldots, \bar{i}_n, \ldots, i_{n+m}}$ are the Legendre polynomial coefficients of different orders, $\bar{i}_k = i_k - i_{k+1}$, and $\hat{f}(\boldsymbol{Z}_X)$ is the ω-order approximation of $f(\boldsymbol{Z}_X)$. The core part $\hat{f}(\boldsymbol{Z}_X)$ is the computation of $L_{\bar{i}_0,\bar{i}_1,\ldots,\bar{i}_n,\ldots,i_{n+m}}$. According to Mustapha and Dimitrakopoulos (2010a, 2010b), these coefficients can be obtained as follows:

$$
\begin{aligned}
L_{\bar{i}_0,\bar{i}_1,\ldots,\bar{i}_n,\ldots,i_{n+m}} &= \sqrt{\frac{2\bar{i}_0 + 1}{2}} \cdots \sqrt{\frac{2i_{n+m} + 1}{2}} \\
&\times \sum_{t_0=0}^{\bar{i}_0} a_{t_0,\bar{i}_0} \cdots \sum_{t_{n+m}=0}^{\bar{i}_{n+m}} a_{t_{n+m},i_{n+m}} \int Z_X^{t_0}(x_0)\ldots Z_Y^{t_{n+m}}(y_m) f(\boldsymbol{Z}_X) dZ_X(x_0) dZ_X(x_1)\ldots dZ_Y(y_m) \\
&= \sqrt{\frac{2\bar{i}_0 + 1}{2}} \cdots \sqrt{\frac{2i_{n+m} + 1}{2}} \sum_{t_0=0}^{\bar{i}_0} a_{t_0,\bar{i}_0} \cdots \sum_{t_{n+m}=0}^{\bar{i}_{n+m}} a_{t_{n+m},i_{n+m}} E(Z_X^{t_0}(x_0)\ldots Z_Y^{t_{n+m}}(y_m)),
\end{aligned}
$$

where $E(Z_X^{t_0}(x_0)\ldots Z_Y^{t_{n+m}}(y_m)) := m_{t_0,\ldots,t_{n+m}}^{Z_X}$ is the expression of the cross-variable higher-order moment discussed previously. Its empirical version, namely the cross-variable higher-order statistics, are used to compute $L_{\bar{i}_0,\bar{i}_1,\ldots,\bar{i}_n,\ldots,i_{n+m}}$ and therefore $\hat{f}(\boldsymbol{Z}_X)$ can be obtained.

Based on $\hat{f}(\boldsymbol{Z}_X)$, the CPDF at the reference location x_0 is computed as follows:

$$
\hat{f}_{Z_X}(Z_X(x_0)) = \hat{f}(Z_X(x_0)|Z_X(x_1),\ldots,Z_X(x_n),Z_Y(y_1),\ldots,Z_Y(y_m)) = \frac{\hat{f}(\boldsymbol{Z}_X)}{\int \hat{f}(\boldsymbol{Z}_X) dZ_X(x_0)}
$$

$\hat{f}_{Z_X}(Z_X(x_0))$ provides full information about the inference of $Z_X(\cdot)$ at x_0. For example:

Mean: $\int Z_X(x_0)\hat{f}_{Z_X}(Z_X(x_0)) dZ_X(x_0)$;
Variance: $\int Z_X^2(x_0)\hat{f}_{Z_X}(Z_X(x_0)) dZ_X(x_0) - (E[Z_X(x_0)])^2$;
Percentiles: $\hat{F}_{Z_X}^{-1}(\alpha)$, where $\alpha \in [0,1]$, $\hat{F}_{Z_X}(\cdot)$ is the cumulative distribution function (CDF) obtained from $\hat{f}_{Z_X}(Z_X(x_0))$, and $\hat{F}_{Z_X}^{-1}(\cdot)$ is the inverse function of $\hat{F}_{Z_X}(\cdot)$.

For demonstration purposes we considered one secondary variable only. The validity of the HOS method remains if there are two or more secondary variables, as the cross-variable higher-order statistics can be generalized in a straightforward manner.

5.3 MATLAB functions for the implementation of the HOS method

The algorithm for computing $\hat{f}(\boldsymbol{Z}_X)$ can be summarized as follows:

1. Define the reference location x_0;
2. Define the spatial template for x_0 by searching within its neighbourhood;

3. Carry out the exhaustive searching process described in Section 5.2.2;
4. Compute the cross-variable higher-order statistics;
5. Compute the coefficients of Legendre polynomials of different orders;
6. Compute the approximated CPDF $\hat{f}(\mathbf{Z}_X)$ using the results of Step (4) and (5).

Step (1) is for the user to decide and hence no computation function is required. Step (2) and (3) are related to the determination of spatial template and the implementation of the searching process, respectively. MATLAB functions "temp" and "tempsearch" were designed for these two tasks. The implementation of Step (4) is enabled by the function "HOS", which requires the output from "temp" and "tempsearch". In Step (5), the function "L" is used to compute the Legendre coefficients of different orders, and these computed coefficients are then employed by the function "pdfx" to carry out the task in Step (6).

5.3.1 Spatial template and searching process

To implement Step (2) of the algorithm, the following "temp" function was developed:

```
function distance = temp(x0coor, data, datacoor, N)
% This function defines the searching template associated to the reference
location x0
%
% Input
% x0coor: 1-by-3 vector, x, y, z coordinates of x0
% datacoor: nd-by-3 vector, where nd is the number of data points
% N: a pre-specified number of directions of the template.
%
% Output
% distance: N-by-(4+nvar) vector, where nvar is the number of variables

x0coor = kron(ones(size(datacoor,1),1),x0coor);
distance = x0coor - datacoor;
distance = [distance sum((distance).^2,2) data];
distance = sortrows(distance, 4);

if distance(1,4) == 0
  distance = distance(2:N+1,:);
else
  distance = distance(1:N,:);
end

end % End of function
```

The following "tempsearch" function carries out the exhaustive searching throughout the entire dataset, that is, the implementation of Step (3):

```
function dataID = tempsearch(datacoor, Ncoor, Ntol)
% This function searches all data points that satisfy the template
%
```

```
% Input
% datacoor: nd-by-3 matrix, x, y, z coordinates of the data set
% Ncoor: N-by-3 matrix, x, y and z coordinates of N template directions
% Ntol: N-by-3 matrix, tolerance for each direction of each template
direction. Must be non-negative.
%
% Output
% dataID: Nd-by-(N + 1) matrix, each row represents a collection of N + 1
data values (including the reference point) that satisfies the template
defined by Ncoor

order = size(Ncoor, 1);  % order of template (number of directions)
nd = size(datacoor, 1);  % number of data points
lower = Ncoor − Ntol;
upper = Ncoor + Ntol;

switch order
% We only show the code for the cases where order = 1 and 2. For higher values
of order, the code can be generalized easily
    case 1
        dataID = zeros(1,2);
        for i = 1:nd
            x0 = datacoor(i,:);
            x0 = kron(ones(nd,1),x0);
            diff = x0 - datacoor;

            [r,~] = find(diff(:,1) > = lower(1,1) & diff(:,1) < = upper(1,1) &
            diff(:,2) > = lower(1,2) & diff(:,2) < = upper(1,2) & diff(:,3) > =
            lower(1,3) & diff(:,3) < = upper(1,3));

            if size(r,1) == 0
                continue;
            else
                new = [ones(size(r))*i r];
                dataID = [dataID; new];
            end
        end

    case 2
        dataID = zeros(1,3);
        for i = 1:nd
            x0 = datacoor(i,:);
            x0 = kron(ones(nd,1),x0);
            diff = x0 - datacoor;

            [r1,~]  =  find(diff(:,1) > = lower(1,1)  &  diff(:,1) < = upper
            (1,1) & diff(:,2) > = lower(1,2) & diff(:,2) < = upper(1,2) & diff
            (:,3) > = lower(1,3) & diff(:,3) < = upper(1,3));
            [r2,~]  =  find(diff(:,1) > = lower(2,1)  &  diff(:,1) < = upper
            (2,1) & diff(:,2) > = lower(2,2) & diff(:,2) < = upper(2,2) & diff
            (:,3) > = lower(2,3) & diff(:,3) < = upper(2,3));
```

```
          s1 = size(r1,1);
          s2 = size(r2,1);

          if s1*s2 == 0
               continue;
          else
               r1 = kron(r1,ones(s2,1));
               r2 = kron(ones(s1,1),r2);

               new = [ones(s1*s2,1)*i r1 r2];
               dataID = [dataID; new];
          end
     end

     % Generalize the code above to deal with cases where order > 2
     end

dataID(1,:) = [];
end % End of function
```

5.3.2   Higher-order statistics

The function "HOS" was produced for the computation of cross-variable higher-order statistics, given a defined spatial template:

```
function m = HOS(data, dataID, order, varID)

% This function computes the high-order statistics
%
% Input
% data: nd-by-nvar matrix where nvar denotes the number of variables
% dataID: Nd-by-(N+1) matrix where N denotes the number of directions in
the template. This is given by the function "tempsearch". Each row
corresponds to a collection of N+1 data values that satisfies the template
% order: 1-by-(N+1) row vector, each element specifies the order of moment
associated with the corresponding direction
% varID: 1-by-(N+1) row vector, indicating which variables are considered
in the template for computing HOS
%
% Output
% m: computed high-order statistics

if size(dataID,2) ~= size(order,2)
     warning('The order vector is not correctly specified')
end

if size(dataID,2) ~= size(varID,2)
     warning('The varID vector is not correctly specified')
end

MM = zeros(size(dataID));
NN = size(dataID,2);
```

```
for i = 1:NN
    MM(:,i) = data(dataID(:,i), varID(i));
end

order = kron(ones(size(dataID,1),1), order);

MM = MM.^order;
MM = prod(MM,2);

m = mean(MM);
end % End of function
```

Now Step (4) in the computing algorithm is completed.

5.3.3   Coefficients of Legendre polynomials

Based on the HOS output from the previous function, the "L" function is employed for the computation of the Legendre coefficients. The details are as follows:

```
function lcoef = L(a, order, data, dataID, varID)

% This function computes the Legendre coefficients of different order
%
% Input
% a: Legendre coefficient matrix, pre-defined
% order: 1-by-(N+1) row vector, where N is the number of directions in the
% searching template. It specifies the orders of the Legendre coefficients.
% e.g. L_{3,4,2,2,0,4,2,3}.
% data: nd-by-nvar matrix where nvar denotes the number of variables
% dataID: Nd-by-(N+1) matrix where N denotes the number of directions in
% the template. This is given by the function "tempsearch"
% varID: 1-by-(N+1) row vector, indicating which variables are considered
% in the template
%
% Output
% lcoef: computed Legendre coefficient, a scalar

% Initial value of the output
lcoef = 0;

oo = size(order, 2);

switch oo
% We only show the code for the cases where oo = 1 and 2. For higher values of
% oo, the code can be generalized easily
    case 1
        for i1 = 0:order(1)
            order0 = i1;
            lcoef0 = a(i1+1,order(1)+1)*HOS(data, dataID, order0, varID);
            lcoef = lcoef + lcoef0;
        end
```

```
    case 2
      for i1 = 0:order(1)
          for i2 = 0:order(2)
              order0 = [i1 i2];
              lcoef0  =  a(i1+1,order(1)+1)  *a(i2+1,order(2)+1)  *HOS
              (data,dataID,order0,varID);
              lcoef = lcoef + lcoef0;
          end
      end

    % Generalize the code above to deal with cases where oo > 2
    end

    % Normalize the coefficient
    lcoef = prod(sqrt((2*order+1)/2))*lcoef;

    end % End of function
```

Step (5) is now completed.

5.3.4   CPDF approximation

Given the computed cross-variable higher-order statistics and the Legendre polynomial coefficients, the "pdfx" function is used to approximate the CPDF at x_0:

```
function pdfx0 = pdfx(x0, x0coor, varID, datacoor, data, Ntol, omega)

% This function approximates the CPDF at the reference location x0
%
% Input
% x0: a user-specified grid over [-1, 1], over which the density function is
% evaluated
% x0coor: 1-by-3 vector, x, y, z coordinates of x0
% varID: 1-by-(N+1) row vector, indicating which variables are considered
% to form the template when computing higher-order statistics. The first
% element specifies the primary variable.
% datacoor: nd-by-3 vector, where nd is the number of data points
% data: nd-by-nvar matrix where nvar denotes the number of variables
% Ntol: N-by-3 matrix, tolerance for each direction of the template
% omega: a scalar, the maximum order to which the pdf is approximated
%
% Output
% pdfx0: the approximated probability density function at x0

% Define the number of direction in the template
N = size(varID, 2) - 1;

varid = find(varID == varID(1));
varid(1) = [];
n1 = size(varid,2);
```

```
n2 = N-n1;
n = max(n1, n2);

% Define the N-direction template
distance = temp(x0coor, data, datacoor, n);
dis1 = distance(1:n1,:);
dis2 = distance(1:n2,:);
distance = [dis1; dis2];

Ncoor = distance(:,1:3);

% Define the closest data values of x0, according to "varID"
datanear = distance; datanear(:,1:4) = [];
neardata = zeros(N,1);

for i = 1:N
    neardata(i) = datanear(i, varID(i+1));
end

% Search data pairs that satisfy the template
dataID = tempsearch(datacoor, Ncoor, Ntol);

% Define the Legendre Polynomials
a = zeros(11);
a(1,1) = 1;
a(1,2) = 0;   a(2,2) = 1;
a(1,3) = -0.5; a(2,3) = 0;   a(3,3) = 1.5;
a(1,4) = 0;   a(2,4) = -1.5;   a(3,4) = 0; a(4,4) = 2.5;
% ...
% ... the rest of values in the matrix "a" is omitted here. Refer to Liu and
Spiegel (1999) for further information

% Legendre Polynomial of any order
% LP = @(x) a(1,i) + a(2,i)*x.^1 + a(3,i)*x.^2 + a(4,i)*x.^3 + a(5,i)*x.^4 +
a(6,i)*x.^5 + a(7,i)*x.^6 + a(8,i)*x.^7 + a(9,i)*x.^8 + a(10,i)*x.^9 + a
(11,i)*x.^10;

% Compute the conditional pdf
pdfx0 = zeros(size(x0));

switch N
% We only show the code for the cases where N = 1 and 2. For higher values of
N, the code can be generalized easily
  case 1
    for i0 = 0:omega
        for i1 = 0:i0

    Lorder = [i0-i1 i1];
    Lcoef = L(a, Lorder, data, dataID, varID);
    LP = 1;
    for j = 1:N
        ooo = Lorder(j+1);
        xx = neardata(j);
```

```
        LPj = a(1,ooo+1) + a(2,ooo+1)*xx.^1 + a(3,ooo+1)*xx.^2 + a(4,
        ooo+1)*xx.^3  +  a(5,ooo+1)*xx.^4  +  a(6,ooo+1)*xx.^5  +  a(7,
        ooo+1)*xx.^6  +  a(8,ooo+1)*xx.^7  +  a(9,ooo+1)*xx.^8  +  a(10,
        ooo+1)*xx.^9+ a(11,ooo+1)*xx.^10;
        LP = LP*LPj;
      end
      pdfx00 = a(1,i0-i1+1) + a(2,i0-i1+1)*x0.^1 + a(3,i0-i1+1)*x0.^2
      + a(4,i0-i1+1)*x0.^3 + a(5,i0-i1+1)*x0.^4 + a(6,i0-i1+1)*x0.^5
      + a(7,i0-i1+1)*x0.^6 + a(8,i0-i1+1)*x0.^7 + a(9,i0-i1+1)*x0.^8
      + a(10,i0-i1+1)*x0.^9+ a(11,i0-i1+1)*x0.^10;
      pdfx00 = pdfx00*Lcoef*LP;
      pdfx00 = prod(sqrt((2*Lorder+1)/2))*pdfx00;
      pdfx0 = pdfx0+ pdfx00;

      end
   end


case 2
   for i0 = 0:omega
     for i1 = 0:i0
        for i2 = 0:i1
     Lorder = [i0-i1 i1-i2 i2];
     Lcoef = L(a, Lorder, data, dataID, varID);
     LP = 1;
     for j = 1:N
       ooo = Lorder(j+1);
       xx = neardata(j);
       LPj = a(1,ooo+1) + a(2,ooo+1)*xx.^1 + a(3,ooo+1)*xx.^2 + a(4,
       ooo+1)*xx.^3  +  a(5,ooo+1)*xx.^4  +  a(6,ooo+1)*xx.^5  +  a(7,
       ooo+1)*xx.^6  +  a(8,ooo+1)*xx.^7  +  a(9,ooo+1)*xx.^8  +  a(10,
       ooo+1)*xx.^9+ a(11,ooo+1)*xx.^10;
       LP = LP*LPj;
     end
     pdfx00 = a(1,i0-i1+1) + a(2,i0-i1+1)*x0.^1 + a(3,i0-i1+1)*x0.^2
     + a(4,i0-i1+1)*x0.^3 + a(5,i0-i1+1)*x0.^4 + a(6,i0-i1+1)*x0.^5
     + a(7,i0-i1+1)*x0.^6 + a(8,i0-i1+1)*x0.^7 + a(9,i0-i1+1)*x0.^8
     + a(10,i0-i1+1)*x0.^9+ a(11,i0-i1+1)*x0.^10;
     pdfx00 = pdfx00*Lcoef*LP;
     pdfx00 = prod(sqrt((2*Lorder+1)/2))*pdfx00;
     pdfx0 = pdfx0+ pdfx00;

        end
      end
   end

% Generalize the code above to deal with cases where N > 2
end

% Normalize the estimated pdf
```

```
constant = (max(x0)-min(x0))/(size(x0,1)-1);
constant = sum(pdfx0*constant);
pdfx0 = pdfx0/constant;

end % End of function
```

The column vector "pdfx0" is the final output of the computation algorithm, consisting of probability density values over the predefined grid. The functions provided above suffice for the approximation of CPDF at any user-specified location. Note that these functions were developed using MATLAB 2012b, whereas the use of particular in-built functions may have changed in the latest version.

5.4 A case study

We consider the application of the HOS method in an open-pit mining project.[1] In mining projects, samples are collected by drilling into the orebody. These samples are therefore referred to as drill-hole samples, which provide information about characteristics of orebodies. The figure below displays the distribution of typical drill holes in a mining project, consisting of about 1 million sampling locations in the 3D space (Figure 5.1).

Due to the high cost of drilling, samples of orebodies can be of the order of one part in a million. As a consequence, spatial statistical techniques are widely applied to the datasets obtained from drill-hole samples so that one can evaluate, for example, the properties of spatially distributed random variables that are of interest (e.g., metal grade). To illustrate, we apply the HOS method to the observed data of two random

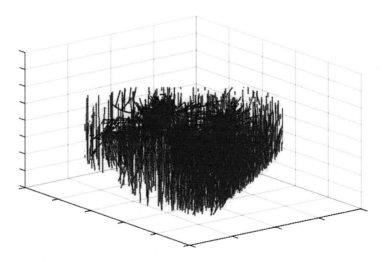

Figure 5.1 A 3D plot of the drill-hole data.

---

[1] Due to confidentiality issues we do not provide details of this mining project.

variables: copper grade and sulphur content. Copper grade (Cu) is considered as the primary variable while sulphur content (S) serves as a secondary variable of Cu. The following computation is carried out:

```
% Read data
data = xlsread('XXXX.xlsx');
% "data" consists of five columns. The first three columns are coordinates
while Columns 4 and 5 are Cu and S values, respectively

% Specify primary and secondary variables
v1ID = 4; % This is the primary variable (in column 4)
v2ID = 5; % This is the secondary variable (in column 5)

% Determine the order of approximation
omega = 8; % The approximation of the CPDF is to the 8th order

% Specify the template
nV1 = 4; % 4 directions (arms) in the template are associated to the
primary variable
nV2 = 2; % 2 directions (arms) in the template are associated to the
secondary variable
% The two commands above jointly imply a 6-direction template

varID = [ones(1,nV1+1),2*ones(1,nV2)];
N = size(varID,2)-1;
% The two commands above follow the specification of the template

% Set tolerance
tol = 0.2; % This is user-specified
Ntol = ones(N,3)*tol; % This setting imposes the same tolerance on the
three dimensions x, y and z. One may wish to impose different levels of
tolerance on x, y and z

% Coordinates of data
datacoor = data(:,1:3);
% Sample of primary and secondary variables
sample = data(:,[v1ID v2ID]);

interval = 0.001;
x0 = (-1:interval:1)'; % The grid over [-1, 1] is defined. A finer or
coarser grid (i.e. a smaller or larger "interval") is also acceptable

% Specify the location at which the CPDF is going to be approximated
x0coor = [xc, yc, zc] % xc, yc and zc are coordinates on the x, y and z
dimensions, respectively

% Implement the HOS method by the "pdfx" function
pdfx0 = pdfx(x0, x0coor, varID, datacoor, sample, Ntol, omega);
% The CPDF is now available for further inference
```

A brief discussion of some inputs is provided below:

- "omega": theoretically a higher omega value will lead to better approximation. However, Liu et al. (2014) argued that in practice, setting omega at 4, 6, 8 or 10 does not make much difference.

- "nV1" and "nV2": these two inputs specify the complexity of the searching template. nV1 + nV2 is the number of directions (arms) in the searching template, where nV1 of them correspond to the primary variable and the rest correspond to the secondary variable. nV1 cannot be zero, whereas nV2 = 0 implies that single-variable statistics will be computed. To some extent, the ratio of nV2 to nV1 reflects the importance of the secondary variable.
- "tol": this is a parameter that has impact on (i) how well the locations of searched data values agree with the template, and (ii) how many data values that satisfy the template can be found. There is a trade-off between (i) and (ii): a smaller "tol" value indicates better matching between the template and data points, but it also implies a smaller proportion of data values that would be identified as "replicates" of the template. We recommend considering relatively small tolerance if the sample size is sufficiently large, or the coordinates of data are fairly regularly spaced.
- "interval": it has impact on the "resolution" of the grid over which the CPDF is evaluated. Preliminary experiments showed that interval ≤ 0.005 is sufficient.

The approximated CPDF, recorded in the vector "pdfx0", can be used to make inferences about the copper grade at the user-specified location with the information about sulphur content taken into account. Examples are:

```
% The expected value
Ex0 = sum(pdfx0.*x0)*interval;

% Cumulative distribution function CDF
cdfx0 = cumsum(pdfx0)*interval;

% Prediction interval
alpha = 0.05 % Level of significance
[~, index] = min(abs(cdfx0-alpha/2));
LB = x0(index); % Lower bound of the 95% prediction interval
[~, index] = min(abs(cdfx0-(1-alpha/2)));
UB = x0(index); % Upper bound of the 95% prediction interval
```

By changing the input "x0coor", the user is able to make inferences about the value of primary variable at any location within the study region. Liu et al. (2014) demonstrated the reliability of the HOS method in the context of spatial interpolation. They concluded that (i) compared to single-variable statistics, the cross-variable HOS noticeably improved the quality of spatial interpolation; (ii) the HOS method performed well when the number of directions in the searching template was no smaller than four; and (iii) the HOS method outperformed the commonly applied kriging method as being more reliable in uncertainty quantification.

In addition to the findings of Liu et al. (2014), we noticed that the computation of HOS is considerably fast, especially when data were observed at regularly spaced locations. It is for these reasons that we recommend the HOS method presented in this chapter to those who work with massive spatial datasets.

References

Arpat, B., & Caers, J. (2007). Conditional simulation with patterns. *Mathematical Geosciences*, *39*, 177−203.

Babak, O. (2014). Inverse distance interpolation for facies modeling. *Stochastic Environmental Research and Risk Assessment, 28,* 1373−1382.

Babak, O., & Deutsch, C. V. (2009). Statistical approach to inverse distance interpolation. *Stochastic Environmental Research and Risk Assessment: Research Journal, 23,* 543−553.

Boucher, A. (2009). Considering complex training images with search tree partitioning. *Computers and Geosciences, 35,* 1151−1158.

Dimitrakopoulos, R., Mustapha, H., & Gloaguen, E. (2010). High-order statistics of spatial random fields: Exploring spatial cumulants for modelling complex, non-Gaussian and non-linear phenomena. *Mathematical Geosciences, 42,* 65−99.

Cressie, N. A. C. (1993). *Statistics for spatial data.* New York: Wiley.

Goovaerts, P. (1997). *Geostatistics for natural resources evaluation.* New York: Oxford University Press.

Hwang, Y., Clark, M., Rajagopalan, B., & Leavesley, G. (2012). Spatial interpolation schemes of daily precipitation for hydrologic modeling. *Stochastic Environmental Research and Risk Assessment : Research Journal, 26,* 295−320.

Liu, J., & Spiegel, M. R. (1999). *Mathematical handbook of formulas and tables* (2nd ed.). New York: McGraw-Hill.

Liu, S., Anh, V., McGree, J., Kozan, E., & Wolff, R. C. (2014). A new approach to spatial data interpolation using higher-order statistics. *Stochastic Environmental Research and Risk Assessment,* 1−12. Available from http://dx.doi.org/10.1007/s00477-014-0985-1.

Mustapha, H., & Dimitrakopoulos, R. (2010a). A new approach for geological pattern recognition using high-order spatial cumulants. *Computers & Geosciences, 36,* 313−343.

Mustapha, H., & Dimitrakopoulos, R. (2010b). High-order stochastic simulation of complex spatially distributed natural phenomena. *Mathematical Geosciences, 42,* 457−485.

Mustapha, H., & Dimitrakopoulos, R. (2011). HOSIM: A high-order stochastic simulation algorithm for generating three-dimensional complex geological patterns. *Computers & Geosciences, 37,* 1242−1253.

Strebelle, S. (2002). Conditional simulation of complex geological structures using multiple point statistics. *Mathematical Geosciences, 34,* 1−22.

Tobler, W. R. (1970). A computer movie simulating urban growth in the Detroit region. *Economic Geography, 46,* 234−240.

Zhang, T., Switzer, P., & Journel, A. G. (2006). Filter-based classification of training image patterns for spatial simulation. *Mathematical Geosciences, 38,* 63−80.

Big data and design of experiments

J.M. McGree[a], C.C. Drovandi[a], C. Holmes[b], S. Richardson[d], E.G. Ryan[a,c] and K. Mengersen[a]
[a]School of Mathematical Sciences, Queensland University of Technology,
[b]Department of Statistics, University of Oxford,
[c]Biostatistics Department, Institute of Psychiatry, Psychology and Neuroscience, King's College, London,
[d]MRC Biostatistics Unit, Cambridge Institute of Public Health, Cambridge, UK

6.1 Introduction

With the mass of data being collected in areas across health, science, business, economics, ecology and robotics, a large focus of recent research in statistics has been the development of big data analytic tools which facilitate statistical inference in big data settings. Recent reviews of approaches to handle and analyse big data have been given by Fan, Han, and Liu (2014) and Wang, Chen, Schifano, Wu, and Yan (2015) which show that the main thrust of current research is the focus on the development of statistical methods that scale to problems of considerable size. Many of the recently proposed approaches adopt a form of 'divide-and-conquer' or 'divide-and-recombine' approach where subsets of the big data are analysed in parallel, and the results are subsequently combined (Guhaa et al., 2012). The division of the full dataset into subsets provides datasets which are much more manageable, and typically straightforward to analyse in parallel. Guhaa et al. (2012) advocate this approach as being applicable to 'almost any' statistical method for large datasets. Visualization techniques are used to minimize the chance of losing important information (when subsetting the data), and approaches are developed to determine the 'best' way to divide the data and recombine the results. Xi, Chen, Cleveland, and Telkamp (2010) have used such an approach to partition internet traffic data for efficiently building complex statistical models to describe the full data. The work was implemented in R and Hadoop via RHIPE (a merger of both of these computational environments). Hadoop is a distributed database and parallel computing platform which distributes computations across central processing units (CPUs). RHIPE allows one to implement such practices from within R.

Zhang, Duchi, and Wainwright (2012) considered approaches for distributed statistical optimization for large-scale datasets. They considered the average mixture algorithm where estimates from the analysis of distinct subsets of the full data are averaged across all CPUs. Further, they extended this simple averaging idea via the consideration of resampling (or bootstrapping) techniques. Each CPU performs a further subsampling analysis to estimate the bias in its estimate, and returns a subsample-corrected estimate which is subsequently averaged over all CPUs. As an alternative approach, Kleiner, Talwalkar, Sarkar, and Jordan (2014) proposed the 'bag of little bootstraps' methodology where results from bootstrapping multiple subsets are averaged to yield the estimate of interest. This method is promoted as being more robust and computationally efficient than alternative 'divide-and-recombine' methods.

In terms of Bayesian inference, consensus Monte Carlo has been proposed by Huang and Gelman (2005) with a similar approach being adopted recently by Scott, Blocker, and Bonassi (2013). Consensus Monte Carlo works by subsampling the data, sampling from the posterior distributions based on each subsample, and then combining these posterior samples to obtain a consensus posterior for statistical inference. Such an approach seems to provide reasonable results (when compared to the posterior distribution obtained by analyzing the full dataset) for posteriors which are well approximated by a normal distribution. In other cases, this approach has been shown to not provide reasonable results. Other Bayesian inference approaches which are applicable to big data problems are methods which are able to form fast approximations to posterior distributions. Such methods include the integrated nested Laplace approximation (INLA) as given by Rue, Martino, and Chopin (2009) and variational approaches as given by, for example, Jaakkola and Jordan (2000) and Ghahramani and Beal (2001). Both of these approaches are deterministic approximations which generally yield reasonable approximations to the posterior marginal distributions. However, the accuracy of approximating the joint posterior distribution requires further investigation for a wide variety of problems/models. Further, INLA employs numerical techniques which suffer from the curse of dimensionality. For example, INLA uses numerical methods to approximate the marginal likelihood of a model. Hence, current approaches for Bayesian inference in big data settings have limitations which prevent their wide spread use in such settings.

Recent developments of statistical modelling approaches have focussed on methods that scale well with the size of big data. Well known multivariate methods such as principal components analysis (Elgamal & Hefeeda, 2015; Kettaneh, Berglund, & Wold, 2005) and clustering approaches (Bouveyrona and Brunet-Saumard, 2014) have been investigated in terms of their applicability in big data settings. Variable selection is a common problem in statistical modelling, and big data settings present further challenges in terms of the dimensionality of the problem. Correlation learning and nonparametric procedures of Fan and Lv (2008) and Fan, Feng, and Rui Song (2011) are relevant, as is the least angle regression approach of Efron, Hastie, Johnstone, and Tibshirani (2004).

This introduction has shown that the majority of approaches proposed to tackle the analysis of big data are methods which require the consideration of the entire dataset (albeit in some cases through divided or partitioned data approaches). In this chapter, we follow an alternative approach given by Drovandi, Holmes, McGree, Mengersen, and Ryan (2015) that focusses on analysing an informative subset of the full data in order to answer specific analysis questions. This is facilitated via an experimental design approach where the big dataset is treated as the population, and informative data are extracted and analysed. In this way, one avoids applying potentially complex and advanced statistical procedures (that may not scale to big data problems) on the full dataset, and therefore should facilitate appropriate inference in complex big data settings. We thus use an experimental design approach to make the problematic big data analysis more computationally tractable.

This chapter focusses on fully Bayesian design approaches and thus we adopt one of the approaches as given in Drovandi et al. (2015). Here, an initial learning phase is adopted in order to form prior information about the model/s and the parameter values. Given this information, sequential designs are found by optimising a utility function to determine which data should be extracted from the full dataset. Once extracted, all data which have been 'collected' are analyzed, and prior information is updated to reflect all information gained. This iterative procedure continues until a fixed number of data points have been extracted. Such an approach was proposed for big data settings by Drovandi et al. (2015) who also considered locally optimal design approaches of similar purposes.

The key to considering an experimental design approach in big data settings is being explicit about the aim of the analysis through the specification of a utility function. Such aims could be, for example, obtaining precise parameter estimates for a given model or determining the influence of a particular covariate on the response. It is this utility which can be used to compare subsets of the full data, and to extract data which are informative in answering the analysis aim/s. As such, we do not advocate this approach in situations where general data mining procedures would be applied as typically in such circumstances the goal is knowledge discovery rather than information gathering (constructed learning).

Thus, we investigate the use of experimental design for the analysis of big data in this chapter. The procedures for determining which designs to extract from the full dataset are defined as well as relevant utility functions for answering specific analysis aims. We exemplify our methodology through the consideration of two examples where big data are encountered. The first example presents as data on whether individuals have defaulted on their mortgage, and interest lies in determining whether other information, such as an individuals credit card debt, can be used to predict whether they will default or not. The second example considers data on flights operated by major airline companies in the U.S.A., and there is interest in determining whether a more complex model for late arrivals that has been proposed in the literature is required. We proceed with an introduction to experimental design and formally show how we propose to apply design techniques in big data settings. This is followed by a section for each example which provides motivation for our research. This chapter concludes with some final words, and directions for future research.

6.2   Overview of experimental design

6.2.1   Background

Experimental design is a guided process of collecting data as effectively as possible to inform the aim/s of an experiment. Such aims could include parameter estimation, model discrimination, prediction and combinations thereof. The design process is generally efficient by being economic in the number of observations that are collected.

Much of the early work and subsequent theoretical results in experimental design focussed on linear models. For such cases, it is important to consider uncertainty about the model as deviations from what is assumed can result in relatively little information being collected to inform experimental aims. Such considerations are also relevant when more complex models, such as nonlinear models, are contemplated. Such models have the added difficulty that the performance of a design can depend on the true parameter values. Thus, one needs information about the model/s and corresponding parameter values in order to derive a design which will yield informative data. However, this is typically the reason for conducting the experiment. Thus, prior information can play an important role in experimental design, and can be used in deriving designs that are robust over different models and parameter values.

Sequential design is a particular form of experimental design where data are collected in stages or over time. This allows one to adjust or adapt the experimental design at different stages of the experiment as more information is collected in regards to the aim of the analysis (and the model appropriate for analysis and corresponding parameter values). Thus, sequential designs are particularly useful in nonlinear settings as one can learn about the parameter values over time, and therefore iteratively derive more informative designs. In such settings, it is optimal to consider the next decision in light of the knowledge that future decisions will also be made (backwards induction). However, this is generally too computationally intensive, and alternatively a myopic approach (one-step-ahead) is adopted.

We follow the work of Drovandi et al. (2015) and consider a fully Bayesian design approach for the location of designs with which to extract data from the full dataset. A review of such approaches has been given recently by Ryan, Drovandi, McGree, and Pettitt (2015a). Given a set of models and estimates of the corresponding parameter values, an optimal design can be defined as one that maximizes the expectation of a given utility. That is,

$$d^* = \arg\max_{d \in D} \psi(d), \tag{6.1}$$

where ψ denotes a general utility, D the design region and the expectation is taken with respect to uncertainty about the model appropriate for data analysis $(p(M = m))$, the associated parameter values $(p(\beta_m|m))$ and the prior predictive distribution of the data $(z \sim p(.|m, d))$. Hence, the above expectation can be expressed as follows:

$$\psi(d) = \sum_{m=1}^{K} p(M = m) \int_z \psi(d, m, z) p(z|m, d) \mathrm{d}z, \tag{6.2}$$

where $\psi(d, m, z)$ is the utility evaluated for a given model m upon observing data z, and K denotes the number of rival or competing models.

The above expression simplifies to the following when only a single model is under consideration:

$$\psi(d) = \int_z \psi(d, z) p(z|d) \mathrm{d}z. \tag{6.3}$$

The most common utility functions found in the literature are associated with parameter estimation. For example, McGree et al. (2012b), Drovandi, McGree, and Pettitt (2013), Ryan, Drovandi & Pettitt (2015b,c) consider maximizing the inverse of the determinant of the posorior variance-covariance matrix (or maximizing the determinant of the posterior precision matrix). This can be expressed as follows:

$$\psi(d, z) = 1/\det[\text{VAR}(\beta|z, d)]. \tag{6.4}$$

In the work of McGree, Drovandi, White, and Pettitt (2015), the inverse of the sum of the posterior variances of parameters was maximized. This can be expressed as:

$$\psi(d, z) = 1/\sum \text{diag}[\text{VAR}(\beta|z, d)].$$

Finally, the Kullback-Liebler divergence (Kullback & Leibler, 1951) between the prior and posterior has also been considered for estimation, see see Shannon (1948), Lindley (1956), Borth (1975), Drovandi et al. (2013), McGree (2015).

Many of such estimation utilities are based on classical design criteria where functionals of the expected Fisher information matrix are used in evaluating the utility of a design. For example, an A-optimal design is one that minimises the trace of the inverse of this information matrix while a D-optimal design maximises the determinant of this information matrix (Fedorov, 1972, Pukelsheim & Torsney, 1991, Atkinson & Donev, 1992).

Another utility that is used in Bayesian design is the mutual information utility for model discrimination, see Box and Hill (1967). The utility favors designs which maximise the expectation of the logarithm of the posterior model probability. As shown in Drovandi, McGree, and Pettitt (2014), the utility can be express as:

$$\psi(d, z, m) = \log p(m, z|d). \tag{6.5}$$

Alternative classical design criteria for model discrimination include T-optimality, see Atkinson and Fedorov (1975a), Atkinson and Fedorov (1975b) and Atkinson and Cox (1974).

In conducting experiments, it is often not possible to collect data at exactly a specified optimal design. This motivated the development of what are known as sampling windows (Duffull, Graham, Mengersen, & Eccleston, 2012; McGree, Drovandi, & Pettitt, 2012a) which define regions in the design space which (when sampled from) yield highly informative designs. Sampling windows may be useful in big data settings as one may not be able to extract the optimal design exactly but there may be a similarly informative design that could alternatively be extracted. Approaches for deriving sampling windows under a fully Bayesian design framework currently do not appear in the literature, but could be pursued as further research into the future.

6.2.2 Methodology

The principle with which Drovandi et al. (2015) propose to apply experimental design techniques for the analysis of big data is straightforward. Utility functions are derived that quantitatively allow designs to be compared in terms of providing information about the aim of the analysis. These could include, for example, estimating parameters in a model to determine if a given effect size is greater than some value of interest, and/or providing precise but independent estimates of parameters of interest. Once the utility has been constructed, it is generally beneficial to determine what prior information is available for addressing the aim of the analysis. This could take the form of information from prior data which have been collected or expert elicited data. If no such prior information is available, an initial learning phase is proposed where data are extracted from the full dataset.

Approaches to extract these data could include random selection or it could be based on designs which are prevalent in the experimental design literature such as factorial designs, fractional factorial designs, central composite designs (circumscribed or inscribed), latin hypercube designs and Box-Behnken designs. The usefulness of these designs will generally depend upon the analysis plan, for example, what type of model will be fitted to the data and what the aims of the analysis may be. Woods, Lewis, Eccleston, and Russell (2006) empirically evaluated the performance of two-level factorial designs for estimating the parameters of a logistic generalized linear model (GLM), and found that such designs generally yield unsatisfactory results. So some care should be taken in deciding which type of design to use in the initial learning phase to ensure that useful information is extracted from the full dataset, and also so that candidate models are not inadvertently eliminated. For example, if only the high and low levels of the covariates are considered (as is the case for two-level factorial designs), then curvature will not be estimable. Hence, as a consequence of the design, only 'straight-line' relationships will be deemed appropriate, despite this not necessarily being the case. We have included this initial learning phase at the start of our algorithm for sequential design for big data specified in Algorithm 6.1.

The initial learning phase is used to develop prior information (for example, about the model/s and corresponding parameter values) so that sequential design

---

Algorithm 6.1 Sequential design algorithm in big data settings

1. Learning phase: Use initial design to extract X_0 and y_0
2. Remove X_0 and y_0 from the original dataset
3. Find a candidate set of model/s which relate y_0 to X_0
4. Obtain $p(\beta_{m,0}|m, y_0, X_0)$ as prior distributions, for $m = 1, \ldots, K$
5. **For** $t = 1$ to T (where T is the number of iterations)
6.   Find $d^* = \arg\max_{d \in D} \psi(d)$
7.   Extract x from original dataset such that $||x - d^*||$ is minimised
8.   Extract y corresponding to x
9.   Remove x and y from the original dataset.
10.   Update $X_{1:t-1}$ and $y_{1:t-1}$ to yield $X_{1:t}$ and $y_{1:t}$
11.   Obtain $p(\beta_{m,t}|m, y_{1:t}, X_{1:t})$, for $m = 1, \ldots, K$
12. **End**

techniques can be implemented. Based on this information, it is useful to develop prior distributions which capture the information gained. For the parameters, in cases where minimal posterior correlation has been observed, normal distributions for parameters will be constructed. Alternatively, when posterior correlation is observed, a multivariate normal distribution will be constructed. In either case, the priors will be constructed based on the results of maximum likelihood estimation. Further, when model uncertainty is considered, we assume all models are equally likely. Based on this prior, one iteratively searches for the next optimal design conditional on the information that has already been extracted. This iterative process continues until sufficient information has been collected to answer the aim of the analysis or, as shown in Algorithm 6.1, until a fixed number of data points have been extracted (denoted as T). Other stopping rules may be appropriate, including evaluating a loss in efficiency metric which shows how informative the extracted data are compared to the optimal design.

For the particular examples that will be considered in this chapter, we will focus on parameter estimation. As shown above, evaluating such utilities requires evaluating integrals with respect to the parameter values and also the future data. Unfortunately, these integrals will generally be intractable, and hence will need to be approximated. Our approach for approximating these utility functions in a sequential design process is now discussed.

In this big data setting, data will be extracted from the full dataset sequentially. This means, as data are extracted, a sequence of target/posterior distributions will be created which need to be sampled from or approximated. Following the approach and setup of Drovandi et al. (2014), this sequence of target distributions can be described as follows:

$$p(\beta_{m,t}|m, \mathbf{y}_{1:t}, \mathbf{d}_{1:t}) = p(\mathbf{y}_{1:t}|m, \beta_{m,t}, \mathbf{d}_{1:t})p(\beta_{m,t}|m)/p(\mathbf{y}_{1:t}|m, \mathbf{d}_{1:t}), \quad \text{for } t = 1, \ldots, T,$$

where $\mathbf{y}_{1:t}$ denotes data collected up until iteration t at designs $\mathbf{d}_{1:t}$, for $m = 1, \ldots, K$. We note that the above sequence of target distributions considers the optimal designs $\mathbf{d}_{1:t}$. However, in this work, these optimal designs need to be extracted from the big dataset. Hence, the extracted design may not be the optimal design (exactly). As shown in Algorithm 6.1, the extracted designs are those with the smallest Euclidean distance from the optimal, and are denoted as $X_{1:t}$.

As T may potentially be quite large, an efficient computational approach to approximating the above target distributions is required. Drovandi et al. (2015) propose to use the sequential Monte Carlo (SMC) algorithm for this purpose, and we elaborate on their methods. This SMC algorithm is based on the work of Chopin (2002), Del Moral, Doucet, and Jasra (2006), and has been used to facilitate efficient computation in the sequential design of experiments recently (Drovandi et al., 2013, 2014; McGree et al., 2015; McGree, 2015). Suppose at the t th iteration of the sequential design process that we have a set of particles $\{\beta_{m,t}^{(i)}, W_{m,t}^{(i)}\}_{i=1}^{N}$ which are a particle approximation to the target distribution $p(\beta_{m,t}|m, \mathbf{y}_{1:t}, \mathbf{d}_{1:t})$. As new data are extracted from the big dataset, this particle approximation needs to be updated. Here, we assume independent data are encountered and thus update the particle approximation straightforwardly using importance sampling, and denote the $t + 1$th particle set as $\{\beta_{m,t}^{(i)}, W_{m,t+1}^{(i)}\}_{i=1}^N$. However, if the weights of the particle approximation are quite variable, the effective sample size may be undesirably small. In such cases, following the importance sampling step, the particle set is resampled to duplicate promising particles and eliminate less promising particles. This is followed by a move step which diversifies the particle set via a Markov chain Monte Carlo kernel. In such cases, the particle set can be denoted as $\{\beta_{m,t+1}^{(i)}, W_{m,t+1}^{(i)}\}_{i=1}^N$, where $W_{m,t+1}^{(i)} = 1/N$, for $m = 1, \ldots, K$ and $i = 1, \ldots, N$.

The particle filtering approach of the SMC algorithm also allows particle approximations of integrals to be formed. For example, $p(y_{1:t}|m, d_{1:t})$ and hence posterior model probabilities can be approximated by noting that:

$$p(y_{1:t+1}|m, d_{1:t+1})/p(y_{1:t}|m, d_{1:t}) = \int_{\beta_{m,t}} p(y_{t+1}|m, \beta_{m,t}, d_{t+1})p(\beta_{m,t}|m, y_{1:t}, d_{1:t})d\beta_{m,t},$$

which can be approximated as follows:

$$p(y_{1:t+1}|m, d_{1:t+1})/p(y_{1:t}|m, d_{1:t}) \approx \sum_{i=1}^N W_{m,t}^{(i)} p(y_{t+1}|m, \beta_{m,t}^{(i)}, d_{t+1}).$$

This is also equivalent to the predictive probability $p(y_{t+1}|m, y_{1:t}, d_{t+1})$, and hence can be approximated in a similar manner. For further details, the reader is referred to the work of Chopin (2002), Del Moral et al. (2006).

The above particle approximation for each of the t target distributions can be used to efficiently approximate utility functions as given in, for example, Equation (6.2). Further, as only binary response models will be considered in the next section, Equation (6.2) can be simplified as follows:

$$\psi(d) = \sum_{m=1}^K p(m) \sum_{z = \{0,1\}} p(z|m, d)\psi(d, m, z),$$

where z represents supposed new data.

Hence, in a sequential design context, the expected utility of a given design d for the observation of y_{t+1} can be approximated as follows:

$$\sum_{m=1}^K \hat{p}(m|y_{1:t}, d_{1:t}) \sum_{z = \{0,1\}} \hat{p}(z|m, d, y_{1:t}, d_{1:t})\psi(d, m, z).$$

The following two examples are investigated to demonstrate our proposed methodology. In both examples, logistic GLMs (McCullagh & Nelder, 1989) were used to model the data with respect to the influence of four covariates. In each example, a discrete set of possible design points was created, and the optimal design was found via an exhaustive search. In the first case, we focus on parameter estimation (of all model parameters) as our analysis aim while in the second case estimation of a subset of parameters is considered. We note that the methodology we propose is not limited to logistic regression nor these utility functions, but in fact could be applied much more broadly to a wide range of models and analysis aims.

6.3 Mortgage Default Example

For this first example, we consider modelling mortgage default as a function of characteristics of the individuals financial status and also characteristics of the property they wish to take a mortgage out on. This example was also considered in Drovandi et al. (2015) under a fully Bayesian design framework. The aim of their work was to determine which covariates influence the probability of defaulting on the mortgage. As such, they considered a candidate set of competing models, and used a model discrimination utility found in Drovandi et al. (2014) to determine which covariates were important. They considered a simulated mortgage defaults dataset which can be found here: http://packages.revolutionanalytics.com/datasets/. We consider the same dataset for the work presented in this chapter but assume that parameter estimation is the aim of the analysis, and therefore consider the utility given in Equation (6.4). The dataset contains a binary variable indicating whether or not the mortgage holder defaulted on their loan, and data on the following covariates which were normalized so that each have a mean of 0 and a variance of 1:

- Credit score: a credit rating (x_1);
- House age: the age (in years) of the house (x_2);
- Years employed: the number of years the mortgage holder has been employed at their current job (x_3);
- Credit card debt: the amount of credit card debt (x_4); and
- Year: the year the data were collected.

Predicting whether an individual will default on their mortgage is obviously quite useful for lenders. It is of interest to determine the major factor/s contributing to defaulting on a mortgage. As such, the following logistic regression model was considered:

$$\log\left(\frac{p}{1-p}\right) = \beta_0 + \beta_1 x_1 + \beta_2 x_2 + \beta_3 x_3 + \beta_4 x_4,$$

where p is the probability of defaulting on the mortgage.

Hence, design techniques will be implemented to learn about the model parameters to inform the lender about the major factor/s contributing to defaulting on a mortgage. Of interest and for means of comparison, the results shown in Table 6.1 are those obtained by analyzing the full dataset which contains 1,000,000

Table 6.1 Results from analysing the full dataset for year 2000 for the Mortgage default example

Parameter	Estimate	Standard error	p-value
β_0	−11.40	0.12	<0.0001
β_1	−0.42	0.03	<0.0001
β_2	0.20	0.03	<0.0001
β_3	−0.63	0.03	<0.0001
β_4	3.03	0.05	<0.0001

observations. However, in many cases, it may not be possible to obtain such results without implementing complex statistical approaches involving parallel computational architectures. Further, such a dataset may have unknown gaps which could potentially limit the inferences which can be drawn. We investigate the usefulness of our designed approach to the analysis of these big data.

From Table 6.1 we can see that all variables are significant, even the number of years the individual has been employed, despite the relatively small effect size (effect sizes can be compared here as the covariates were scaled to have a mean of 0 and a variance of 1). Indeed, this is a feature of big data. With such a mass of data available, typically most effect sizes are significant. However, there are analysis aims where one need not analyse the full data, for example, determining the most influential factor and the precise estimation of the parameters. This is explored in this example.

In order to undertake Bayesian sequential design in this big data context, we need prior information about the parameter values (and model/s). As we assume no such information is available for this mortgage case study, we adopt an initial learning phase where 10,000 data/design points are selected at random for extraction from the mortgage dataset. At this point, the goodness-of-fit of a variety of models should be investigated to determine a candidate set of models which may be appropriate for data analysis. For this work, we assume that the main effects model with all four covariates is appropriate, and run our sequential design algorithm to iteratively extract additional data from the mortgage dataset for the precise estimation of model parameters (using the criterion given in Equation (6.4)). At each iteration, we assume that all combinations of the covariate levels shown in Table 6.2 were available for selection, resulting in a total of 9900 potential designs.

Table 6.3 shows the results for when the process was stopped after 10,000 observations. The results show that the estimated probability that each effect size is greater than 0 is either very small or very large, hence all covariates influence the chance of defaulting on a mortgage. Overall, the strongest influence on mortgage default appears to be credit card debt with an estimated effect size around 6 to 7 times larger than the next largest.

From the results in Table 6.3, the parameter values obtained by analyzing all 1,000,000 observations are all contained in the 95% credible intervals for each parameter. Further, the posterior standard deviations obtained here are reasonably

Table 6.2 **The levels of each covariate available for selection in the sequential design process for the mortgage example from Drovandi et al. (2015)**

Covariate	Levels
Credit score	$-4, -3, -2, -1, -0.5, 0, 0.5, 1, 2, 3, 4$
House age	$-3, -2, -1, -0.5, 0, 0.5, 1, 2, 3$
Years employed	$-3, -2, -1, -0.5, 0, 0.5, 1, 2, 3, 4$
Credit card debt	$-3, -2, -1, -0.5, 0, 0.5, 1, 2, 3, 4$

Table 6.3 **Results from analysing the extracted dataset from the initial learning phase and the sequential design process for the Mortgage default example**

Parameter	Posterior mean	Posterior SD	Prob. > 0
β_0	-11.06	0.46	0.0000
β_1	-0.28	0.08	0.0020
β_2	0.16	0.08	0.9930
β_3	-0.42	0.08	0.0000
β_4	2.83	0.16	1.0000

small, particularly when compared to the standard errors obtained through considering the whole dataset.

Figure 6.1 shows the 95% credible intervals for all effects at each iteration of the sequential design process. Throughout the whole sequential design process, the credible intervals contain the estimated parameter estimates obtained by analyzing the full dataset. As expected, all intervals are generally becoming smaller as more data are included into the analysis. There appears to be more value in the data included earlier in the process in terms of reducing uncertainty in the estimates when compared to those included later. This may be a result of diminishing returns on including additional data or suggest that the data included later in the analysis are relatively far from the optimal design.

Figure 6.2 shows the optimal designs plotted against the selected designs. Ideally, there should be a one-to-one relationship between the two. However, this is not seen, particularly for credit card debt. Here, when it is desirable to select data on individuals with large credit card debt, only smaller debts were found (in general). Indeed, there are cases where a (scaled) credit card debt of around 5 was desired but individuals with negative values were selected. Hence, there appears to be a gap in the dataset for mortgage defaults for individuals with larger credit card debts. If these were real data (that is, not simulated) then it may not be so surprising that there are few data points on individuals with high credit card debt as such individuals may not have their application for a mortgage approved. Further, there also appears to be a lack of data on individuals with a low credit score. From the plot,

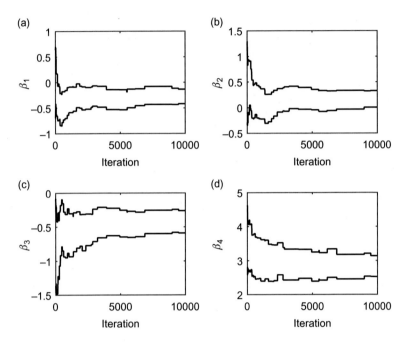

Figure 6.1 95% credible intervals for effect sizes for the Mortgage default example at each iteration of the sequential design process.

when -5 was selected as optimal for x_1, values above 0 were extracted from the full dataset. Again, in reality it would make sense for individuals with low credit scores to not have their mortgage application approved. Lastly, similar statements can be made about house age and years employed. However, the lack of data does not appear as severe (when compared to x_1 and x_4).

Thus, it appears that, despite there being 1,000,000 data records, there are still potential gaps in the dataset, for example, high values for credit card debit. Such gaps limit the efficiency of information that can be obtained on, in this case, parameter estimates. While it may seem obvious that such gaps in these data would exist, other gaps in other big datasets may not be so obvious. Our experimental design approach was able to identify this as a potential issue for parameter estimation. Moreover, we were able to obtain reasonable analysis results via considering less than 1% of the full dataset.

6.4   U.S.A domestic Flight Performance − Airline Example

In this next example, we explore the publically available 'on-time performance of domestic flights' dataset as published by the U.S. Department of Transportation's Bureau of Transportation Statistics (found here: http://stat-computing.org/dataexpo/

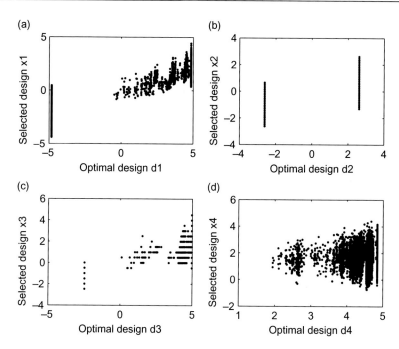

Figure 6.2 Optimal design versus selected design points for (a) credit score, (b) house age, (c) years employed and (d) credit card debt for the Mortgage default example.

2009/the-data.html). The dataset has information on the year, month, day, departure time, origin, destination, flight distance, and many other relevant variables for flights operated by large air carriers for flights from 1987 to 2008. This dataset contains over 160 million records, and has been used to exemplify a range of big data analytic tools by a number of authors, for example, Wang et al. (2015). We follow these authors' approach for using the logistic regression model to quantify the influence of particular covariates on late arrival. Following their work, we created a binary response variable, denoting late arrival, defined as a plane arriving more than 15 minutes after the scheduled arrival time ($y = 1$) or not ($y = 0$). Two continuous covariates were investigated as potential influences of late arrivals: departure hour (x_1, DepHour, range 0 to 24) and distance from origin to destination (x_2 in 1000 miles). Further, two binary covariates were also considered, namely departure time ($x_3 = 1$ if departure occurred between 8pm and 5am and 0 otherwise) and weekend ($x_4 = 1$ if departure occurred on weekends and 0 otherwise). For the purposes of this work, we focus on the year 1995 which has a total of 5,229,619 observations.

This example was also considered by Drovandi et al. (2015). In their work, they were interested in determining whether a more complex model for the data is appropriate than the full main effects model proposed by Wang et al. (2015). To investigate this, the model discrimination utility of Drovandi et al. (2014) was used to determine which model was preferred for data analysis between the full main effects model (Wang et al., 2015) and this full main effects model with an additional quadratic term for x_1 (departure hour). For the work considered in this chapter, we propose the same analysis aim. However, as the discrimination problem is between a larger and a reduced model, we approach optimal design selection via a form of Ds-optimality. Such an approach for discriminating between competing nested regression models has been considered in the literature under a classic design framework, for example, see Dette and Titoff (2009).

The form of the Ds-optimality implemented in this work is a special case of the utility given in Equation (6.4). Instead of focussing on all of the parameters, only a subset is considered. The specific subset of parameters with which the posterior precision will be maximized are those which are additional to the nested model. The classical design literature gives a slightly different formulation of the utility, see Whittle (1973). However, we maintain our focus on maximising posterior precision of the subset of parameters.

As in the previous example, we assume that no prior information about how or which covariates influence the chance of late arrivals is available, and hence use an initial learning phase to develop prior information. Here, we select 10,000 data/design points from the airline dataset at random. Multivariate normal distributions were then constructed based on a maximum likelihood fit to these data for each model. Specifically, the maximum likelihood estimate was set as the prior mean, and the inverse of the observed Fisher information matrix was set as the variance-covariance matrix. Once constructed, the sequential design process was run for the extraction of an additional 5,000 data points. Employing a type of Ds-optimality, an exhaustive search of all combinations of covariate levels given in Table 6.4 was undertaken to determine which design was optimal. Given the levels of the optimal choice of x_3 and x_4, the full dataset was subsetted, and the design within this subset with the smallest Euclidean distance from the optimal was selected for extraction. At each iteration of the sequential design process, the posterior model probability of each model was recorded, along with the optimal design, the extracted design and the Euclidean distance between these two designs.

Figure 6.3 shows the posterior model probabilities at each iteration of the sequential design process for each model in the candidate set. Only the first 1,000 iterations of the 5,000 are shown in the plot as it appears as though the Ds-optimal design approach for discrimination performs well by selecting the more complex

Table 6.4 **The levels of each covariate available for selection in the sequential design process for the airline example from Drovandi et al. (2015)**

Covariate	Levels
x_1	$-3, -2, -1, -0.5, 0, 0.5, 1, 2, 3$
x_2	$-1, -0.5, 0, 0.5, 1, 2, 3, 4$
x_3	0,1
x_4	0,1

model as being preferred for data analysis after around 500 iterations. These results are similar to that given in Drovandi et al. (2015) for the mutual information utility.

To further explore the results from the sequential design process, the distance from the optimal design against the selected design points was inspected (see Figure 6.4). From the plot, it appears as though sufficient data are available for x_2 in order to determine which model is preferred for data analysis. However, for x_1, when extreme positive values were selected as optimal, values much less than these extremes were extracted from the full dataset. For example, when x_1 around 3 was selected as optimal, only values around 1 and 2 were available in the full dataset. Hence, it appears as though there is a lack of data on late night departures which may cause inefficiencies in determining whether a more complex model is required.

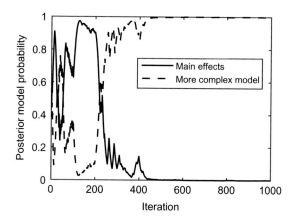

Figure 6.3 Posterior model probabilities at each iteration of the sequential design process for the two models in the candidate set for the airline example.

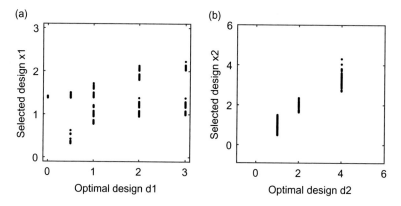

Figure 6.4 Ds-optimal designs versus the designs extracted from the airline dataset for (a) departure hour and (b) distance from the origin to the destination.

As in the previous example, by analysing less than 1% of the available data, we more able to confidently answer our analysis aim and also identify potential gaps the big dataset.

6.5 Conclusion

In this chapter, we investigated the use of experimental design techniques in the analysis of big data. This research was motivated by the ever increasing size of datasets that are currently being collected and therefore the ever increasing need to develop scalable and appropriate statistical methods for big data analytics. The proposed approach, in the absence of prior information, adopts an initial learning phase where data are extracted from the full dataset for analysis. This was followed by a formal sequential design process where specific aims of analysis can be informed by optimal design methods. Under such an approach, analysing the full dataset is avoided. Indeed, the two examples considered in this chapter showed that less than 1% of the full dataset needed to be analysed to inform the experimental aims. This has obvious benefits as one can potentially avoid having to use parallel computational methods and should allow the application of complex statistical methods in situations where it is computationally infeasible to analyse the full dataset.

We stress that do not advocate the use of experimental design techniques for solving all of the analysis issues recently encountered in big data problems. The issue of whether the big data actually represents a given population still remains. Latent unobserved variables could still be influential, and in fact driving the response of interest. A formal designed approach may mitigate against such pitfalls of analysing big data. For example, bias in over representing certain populations can be tackled via a designed approach. Indeed, big data does not have all of the answers, and may not actually be very informative with respect to particular analysis aims. A nice feature of a designed approach is the identification of such gaps, and thus providing a means of evaluating the quality of the collected data.

The designed approach could in fact be considered complementary to the approaches currently employed to analyse big data. More complex statistical approaches could be applied to the extracted dataset while general, more scalable techniques could be applied to the full dataset. Further, big data techniques may benefit from a design approach. For example, design approaches could be employed to divide the full dataset for analysis via the divide-and-conquer approach, the divide-and-recombine approach and consensus Monte Carlo. Moreover, a designed approach could be useful for efficiently dealing with a large volume of streaming data.

Further, our approaches for extracting designs from the big dataset could be improved. Sampling windows seem naturally suited for such endeavours. Adopting such an approach should increase the efficiency of the extracted designs with respect to the optimal design. This could be achieved by extracting a design (or indeed designs) if they appear within the specified sampling window. Further improvements could be achieved by extracting designs with the highest efficiency (with respect to the optimal) rather than extracting the design with the smallest Euclidean distance from the optimal. Such an approach may have been useful in each example as it was seen that there were limited data for a particular covariates. In such cases, it would be more reasonable to consider extracting a design which is highly efficient as there may be a more informative design which is not necessarily the closest (in the Euclidean space) to the optimal.

References

Atkinson, A. C., & Cox, D. (1974). Planning experiments for discriminating between models. *Journal of the Royal Statistical Society: Series B (Statistical Methodology)*, *36*, 321−348.

Atkinson, A. C., & Donev, A. N. (1992). *Optimum experimental designs*. New York: Oxford University Press.

Atkinson, A. C., & Fedorov, V. V. (1975a). The design of experiments for discriminating between two rival models. *Biometrika*, *62*, 57−70.

Atkinson, A. C., & Fedorov, V. V. (1975b). The design of experiments for discriminating between several models. *Biometrika*, *62*, 289−303.

Borth, D. M. (1975). A total entropy criterion for the dual problem of model discrimination and parameter estimation. *Journal of the Royal Statistical Society: Series B (Methodological)*, *37*, 77−87.

Bouveyrona, C., & Brunet-Saumard, C. (2014). Model-based clustering of high-dimensional data: A review. *Computational Statistics & Data Analysis*, *71*, 52−78.

Box, G. E. P., & Hill, W. J. (1967). Discrimination among mechanistic models. *Technometrics*, *9*, 57−71.

Chopin, N. (2002). A sequential particle filter method for static models. *Biometrika*, *89*(3), 539−551.

Del Moral, P., Doucet, A., & Jasra, A. (2006). Sequential Monte Carlo samplers. *Journal of the Royal Statistical Society: Series B (Statistical Methodology)*, *68*(3), 411−436.

Dette, H., & Titoff, S. (2009). Optimal discrimination designs. The. *Annals of Statistics*, *37*, 2056−2082.

Drovandi, C.C., Holmes, C., McGree, J.M., Mengersen, K., Ryan, E.G. (2015). A principled experimental design approach to big data analysis, Available online at <http://eprints.qut.edu.au/87946/>.

Drovandi, C. C., McGree, J. M., & Pettitt, A. N. (2013). Sequential Monte Carlo for Bayesian sequentially designed experiments for discrete data. *Computational Statistics and Data Analysis*, *57*, 320−335.

Drovandi, C. C., McGree, J. M., & Pettitt, A. N. (2014). A sequential Monte Carlo algorithm to incorporate model uncertainty in Bayesian sequential design. *Journal of Computational and Graphical Statistics*, *23*, 3−24.

Duffull, S. B., Graham, G., Mengersen, K., & Eccleston, J. (2012). Evaluation of the pre-posterior distribution of optimized sampling times for the design of pharmacokinetic studies. *Journal of Biopharmaceutical Statistics*, *22*(1), 16−29.

Efron, B., Hastie, T., Johnstone, I., & Tibshirani, R. (2004). Least angle regression. *The Annals of Statistics*, *32*(2), 407−499.

Elgamal, T., Hefeeda, M. (2015). Analysis of PCA algorithms in distributed environments. arXiv:1503.05214v2 [cs.DC] 13 May 2015.

Fan, J., Feng, Y., & Rui Song, R. (2011). Nonparametric independence screening in sparse ultra-high dimensional additive models. *Journal of the American Statistical Association,* *106*(494), 544−557.

Fan, J., Han, F., & Liu, H. (2014). Challenges of big data analysis. *National Science Review,* To appear

Fan, J., & Lv, J. (2008). Sure independence screening for ultrahigh dimensional feature space. *Journal of the Royal Statistical Society: Series B (Statistical Methodology),* *70* (5), 849−911.

Fedorov, V. V. (1972). *Theory of Optimal Experiments.* New York: Academic Press.

Ghahramani, Z., & Beal, M. J. (2001). Propagation algorithms for variational Bayesian learning. *Advances in Neural Information Processing Systems,* 507−513.

Guhaa, S., Hafen, R., Rounds, J., Xia, J., Li, J., Xi, B., & Cleveland, W. (2012). Large complex data: Divide and recombine (D&R) with RHIPE. *Stat, 1,* 53−67.

Huang, Z., Gelman, A. (2005). Sampling for Bayesian computation with large datasets, <http://www.stat.columbia.edu/ gelman/research/unpublished/comp7.pdf>.

Jaakkola, T. S., & Jordan, M. I. (2000). Bayesian parameter estimation via variational methods. *Statistics and Computing, 10,* 25−37.

Kettaneh, N., Berglund, A., & Wold, S. (2005). PCA and PLS with very large data sets. *Computational Statistics & Data Analysis,* *48*(11), 68−85.

Kleiner, A., Talwalkar, A., Sarkar, P., & Jordan, M. (2014). A scalable bootstrap for massive data. *Journal of the Royal Statistical Society: Series B (Statistical Methodology),* *76*(4), 795−816.

Kullback, S., & Leibler, R. A. (1951). On information and sufficiency. *The Annals of Mathematical Statistics, 22,* 79−86.

Lindley, D. (1956). On a measure of the information provided by an experiment. *Annals of Mathematical Statistics, 27,* 986−1005.

McCullagh, P., & Nelder, J. A. (1989). *Generalized linear models* (2nd ed.). Chapman and Hall.

McGree, J., Drovandi, C. C., & Pettitt, A. N. (2012a). A sequential Monte Carlo approach to derive sampling times and windows for population pharmacokinetic studies. *Journal of Pharmacokinetics and Pharmacodynamics,* *39*(5), 519−526.

McGree, J.M. (2015). An entropy-based utility for model discrimination and parameter estimation in Bayesian design, Available online at <http://eprints.qut.edu.au/86673/>.

McGree, J. M., Drovandi, C. C., Thompson, M. H., Eccleston, J. A., Duffull, S. B., Mengersen, K., Pettitt, A. N., & Goggin, T. (2012b). Adaptive Bayesian compound designs for dose finding studies. *Journal of Statistical Planning and Inference, 142,* 1480−1492.

McGree, J.M., Drovandi, C.C., White, G., Pettitt, A.N. (2015). A pseudo-marginal sequential Monte Carlo algorithm for random effects models in Bayesian sequential design. Statistics and Computing. To appear.

Pukelsheim, F., & Torsney, B. (1991). Optimal weights for experimental designs on linearly independent support points. The. *Annals of Statistics, 19*(3), 1614−1625.

Rue, H., Martino, S., & Chopin, N. (2009). Approximate Bayesian inference for latent Gaussian models using integrated nested Laplace approximations (with discussion). *Journal of the Royal Statistical Society, Series B (Statistical Methodology), 71*(2), 319−392.

Ryan, E., Drovandi, C., McGree, J., & Pettitt, A. (2015a). A review of modern computational algorithms for Bayesian optimal design. *International Statistical Review*, Accepted for publication.

Ryan, E., Drovandi, C., & Pettitt, A. (2015b). Fully Bayesian experimental design for Pharmacokinetic studies. *Entropy, 17*, 1063−1089.

Ryan, E., Drovandi, C. C., & Pettitt, A. N. (2015c). Simulation-based fully Bayesian experimental design for mixed effects models. *Computational Statistics & Data Analysis, 92*, 26−39.

Scott, S.L., Blocker, A.W., Bonassi, F.V. (2013). Bayes and big data: The consensus Monte Carlo algorithm. In: Bayes 250.

Shannon, C. (1948). A mathematical theory of communication. *Bell System Technical Journal, 27*, 379−423, 623−656.

Wang, C., Chen, M.-H., Schifano, E., Wu, J., Yan, J. (2015). A survey of statistical methods and computing for big data. ArXiv:1502.07989 [stat.CO].

Whittle, P. (1973). Some general points in the theory of optimal experimental design. *Journal of the Royal Statistical Society: Series B (Statistical Methodology), 35*, 123−130.

Woods, D. C., Lewis, S. M., Eccleston, J. A., & Russell, K. G. (2006). Designs for generalized linear models with several variables and model uncertainty. *Technometrics, 48*, 284−292.

Xi, B., Chen, H., Cleveland, W., & Telkamp, T. (2010). Statistical analysis and modelling of internet VoIP traffic for network engineering. *Electronic Journal of Statistics, 4*, 58−116.

Zhang, Y., Duchi, J.C., Wainwright, M.J. (2012). Communication-efficient algorithms for statistical optimization, <http://arxiv.org/abs/1209.4129>. [stat.ML].

Big data in healthcare applications ⑦

As stated in Chapter 1, big data problems are often characterized by the 5Vs: Volume (quantity of data), Velocity (fast generation of data), Variety (data from different sources and with different structures), Veracity (trustworthy features of data) and Value (the amount of useful knowledge that can be extracted from data). Datasets in the field of healthcare, as well as clinical medicine or health informatics, usually exhibit these characteristics:

- *Volume.* Reports showed that in 2012, the size of healthcare data reached about 500 petabytes, whereas this figure is predicted to be 25,000 petabytes in 2020 (Hermon & Williams, 2014).
- *Variety.* Healthcare data can be either structured (e.g. electronic health records, EMRs) or unstructured (e.g. fMRI images), being collected from various sources.
- *Velocity.* Due to the developments in both hardware and software engineering, a growing amount of EMRs and health insurance data is being generated everyday.
- *Veracity.* Clinical data are generally reliable as most of them are recorded by clinical professionals (e.g. EMRs collected from clinical trials).
- *Value.* Tremendous benefits can be gained if clinical data are used wisely (e.g. to support clinicians carrying out diagnostic, therapeutic or monitoring tasks).

These 5Vs imply the potential of big data analytics for improving healthcare systems and breaking barriers in traditional healthcare and medicine, which can be summarized as follows:

- *'Traditional care and medicine is episodic and symptomatic'* (Pentland, Reid, & Heibeck, 2013). For example, in traditional healthcare, patients usually have to go through a 'black-hole' period between their visits to doctors, during which doctors know little about their health status. As a consequence, only narrow snapshots (i.e. episodic records) can be captured during their visits to doctors whereas most of the other time periods remain blank (Pentland et al., 2013). In addition, traditional public surveillance systems try to monitor patients by indicators like pharmacy purchases, but such indicators are not able to provide timely feedback. In the era of big data, however, with the help of mobile devices which are capable of collecting health data, continuous surveillance of individuals' health status has become possible, which substantially increases the quality of healthcare (Pentland et al., 2013). See Chapter 8 of this book for more details.
- *'Traditional care and medicine is fragmented'* (Viceconti, Hunter, & Hose, 2015). In order to handle the complexity of living organisms, traditional care decomposes them into parts (e.g. cells, tissues, organs, etc.), which are investigated separately by different medical specialists. This can result in difficulties when dealing with multiorgan or systemic diseases (Viceconti et al., 2015). Big data now provides opportunities of integrating information from different levels (genes, molecular, tissue, patient, population), and consequently the interrelationships among these levels can be captured and utilized (Pentland et al., 2013).
- *'Traditional care and medicine has much ambiguity'* (Viceconti et al., 2015). Most traditional clinicians lack a well-established body of knowledge in a quantitative or

Computational and Statistical Methods for Analysing Big Data with Applications.

semiquantitative way (Viceconti et al., 2015), whereas in the era of big data analytic models may help doctors with medical diagnosis, providing more accurate predictions compared to the epidemiology-based phenomenological methods (Viceconti et al., 2015). See Chapter 8 of this book for more details.

The aim of this chapter is to provide the reader with an overview of how the issues in healthcare can be facilitated by data collection, processing and analysis techniques. Various types of healthcare datasets that have been extensively studied are introduced first, followed by a brief description of data mining procedures that are suitable for these datasets. Finally, a case study is provided to demonstrate how big data can contribute to the improvement of healthcare systems.

7.1   Big data in healthcare-related fields

Herland, Khoshgoftaar, and Wald (2014) carried out a comprehensive review of big data analysis in healthcare-related fields. They claimed that big data is being generated from various fields of clinical medicine, including bioinformatics, image informatics (e.g. neuroinformatics), clinical informatics, public health informatics and translational bioinformatics. They also investigated different levels of research questions and data usage, which are reported in Table 7.1.

In this section, we briefly review different categories of healthcare data, as well as how these data can be utilized in a broad range of applications.

7.1.1   Categories of healthcare data

According to Cottle & Hoover (2013) and Weber, Mandl, & Kohane (2014) big data in healthcare can be categorized into the following classes:

- Healthcare centre data: EMRs including diagnosis and procedures, physicians' notes, emails and paper documents.
- Big transaction data: healthcare claims and other billing records.
- Registry or clinical trial data: data format can be diverse, such as X-rays and other medical images, blood pressure, pulse and pulse-oximetry readings, etc.
- Pharmacy data: medication (medication prescribed, allergies, etc.).
- Data outside healthcare systems:
  - Web and social media data: data from social media such as Facebook, Twitter, LinkedIn and smart-phone apps.
  - Lifestyle data: fitness club membership, grocery store purchases, etc.
  - Genetics data: data from genetics websites (e.g. the '23andMe' website).
  - Environment data: climate, weather and public health databases.

Table 7.1 A summary of big data analysis in healthcare

Fields of research	Bioinformatics, Neuroinformatics, Clinical Informatics, Public Health Informatics, Translational Bioinformatics
Research question levels	Human-scale, clinical-scale, epidemic-scale
Data levels	Molecular-level, tissue-level, patient-level, population-level

With huge amounts of data being collected across a broad range of healthcare, there is potential for big data analytics to contribute in various fields, which will be introduced in the next subsection.

7.1.2 Fields of applications

In the literature, numerous studies have been undertaken to extract useful knowledge from various types of healthcare data. The following fields of applications have attracted a large amount of interest:

- *Genomics.* Genomics is the study of complete genetic materials, including sequencing, mapping and analysing RNA and DNA. The volume of data being produced by sequencing, mapping and analysing genomes is huge, thereby fitting well in the big data context. It is one of the fields where big data analysis has been proven extremely successful. A well-known application in genomics is the prediction of genetic risk associated with particular diseases. For instance, BRCA1 and BRCA2 genes are investigated to examine genetic determinants of breast cancer (Akay, 2009).
- *Public health.* The public health surveillance system is now being transformed by data mining methods in clinical medicine. One important application is using EHRs which contain valuable clinical information to increase the efficiency of public health surveillance (Klompas et al., 2012). Algorithms analysing EHR data have been recently developed by physicians at Harvard Medical School (Klompas et al., 2013) to detect and categorize diabetes patients.
- *Mobile sensors and wearable devices.* Mobile sensors and wearable devices are capable of providing timely data and long-lasting measurements. Architectures have been constructed to facilitate sharing resources of health data. For example, the Open mHealth platform aims to integrate data collected by mobile applications so that digital health solutions can be built (Viceconti et al., 2015).
- *Medical imaging.* Medical imaging is one of the areas naturally associated with big data, as high-resolution images are routinely generated during medical examinations (e.g. X-ray radiography, computed tomography (CT) scan, magnetic resonance imaging (MRI), endoscopy, etc.). A recent study was carried out by Smith et al. (2013), who considered the resting-state functional MRI for the purpose of mapping the macroscopic functional connectome to investigate brain networks. They concluded that the functional MRI data can be used to predict fluid intelligence, that is, the capacity of thinking logically and solving problems.
- *Diagnosis and treatment.* The knowledge obtained from big data can be extremely valuable in tailoring medication or treatments for individuals. Successful applications have now reached the clinical assessment stage or even have been implemented in clinical practice. For example, Caroli et al. (2013) developed a patient-specific computational vascular network model that can examine vessel walls and predict blood flow change within 6 weeks after surgery. This innovative approach has enabled the planning of vascular surgery in advance and optimizing surgical procedure, which has the potential to reduce the occurrence of complications.
- *Infectious disease prediction.* Google Flu is one of the most famous big data applications in predicting infectious disease. It detects outbreaks of flu by aggregating Google search queries and counting the number of queries with keywords related to 'flu'. A peak in such queries is a good indicator of an outbreak (Ginsberg et al., 2009).
- *Mental health.* De Choudrhury, Gamon, Counts, and Horvitz (2013) explored the possibility of incorporating social media data in the detection of major depressive disorders. They analysed postings made by a number of Twitter users who were diagnosed with clinical

depression, extracting their behavioural attributes from the posted messages, pictures, videos, etc. It was concluded that social media data are helpful in recognizing mental diseases.

• *Financial issues of healthcare.* Big data analysis has been carried out to solve problems in the financial aspect of healthcare, such as insurance pricing, cost reduction, risk management, fraud detection, hospital operations management, etc. (Bertsimas et al., 2008; Zhao et al., 2005). For example, the Parkland hospital at Dallas, Texas developed a predictive model to identify patients with high-risk in the coronary care unit, while the outcome after discharge can also be predicted. Due to the predictive model, the 30-day readmission rate of those patients with heart failure was reduced by 26.2%, saving approximately $500,000 per year (Amarasingham et al., 2013).

7.1.3 Data mining procedures in healthcare

A standard process for large-scale data analysis was proposed by Peng, Leek, and Caffo (2015), which is displayed by the following figure (Figure 7.1):

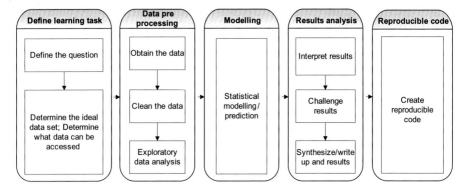

Figure 7.1 Standard process for large-scale data analysis, proposed by Peng, Leek, and Caffo (2015).

In the context of clinical medicine, Bellazzi and Zupan (2008) provided a guideline for predictive data mining. They claimed that predictive data mining methods originate from different research fields and often use very diverse modelling approaches. They stated that the following techniques are some of the most commonly used predictive data mining methods:

• Decision trees
• Logistic regression
• Artificial neural networks
• Support vector machines
• Naive Bayes
• Bayesian networks
• The *k*-nearest neighbours

The reader is referred to Chapter 2 of this book for details of these methods.

Note that according to Bellazzi and Zupan (2008), the methods listed above are often an integral part of modern data mining procedures. For the purpose of model selection, they suggested considering both the predictive performance and interpretability of results. If two methods perform similarly, one should choose the one whose results are easier to interpret.

To implement predictive data mining in clinical medicine, the following tasks should be carried out (Bellazzi & Zupan, 2008):

- *Defining the problem, setting the goals.* At the very beginning, the goal has to be set and the specific problem has to be defined with caution, pointing out the direction of subsequent work. Generally speaking, the aim of predictive data mining in clinical medicine is to make data-informed decisions that help physicians improve their prognosis, diagnosis or treatment planning procedures. To achieve this, preliminary tasks need to be performed. For example, one needs to pay attention to the balance between the predictive power and comprehensibility of a model in the first place, as in particular cases the transparency of data analysis is central to a physician. Decisions also need to be made upon what statistics to use for the assessment of model performance. All these considerations have impact on the subsequent tasks of data mining.
- *Data preparation.* Clinical data usually come from dedicated databases which were purposely collected to study particular clinical problem (e.g. electrocardiogram database for heart disease research). Recently, large-scale data warehouses like hospital records and health insurance claims have been exploited to solve clinical problems. However, due to confidentiality restrictions, many of these databases are extremely difficult to access.
- *Modelling and evaluation.* The objective is to apply a set of candidate methods to the observed clinical data, and to determine which method is most suitable. As mentioned earlier, each data mining method is evaluated by its predictive performance and comprehensibility. While the former is relatively easy to quantify, the latter is a subjective measure that is evaluated by participating domain experts. Note that algorithms working in a black-box manner (e.g. neural network) tend to be not preferred by physicians, even if they may achieve better predictive performance.
- *Deployment and dissemination.* Most clinical data mining projects would terminate once the predictive model has been constructed and evaluated, whereas it is very rare to see reports on the deployment and dissemination of predictive models. One possible reason is that the complexity of data mining tools and the lack of user interface may impede the dissemination in clinical environments. Bellazzi and Zupan (2008) mentioned that the difficulty of deploying and disseminating predictive data mining methods is due to the lack of bridging between data analysis and decision support. Attempts have been made to solve this problem, for example, the Predictive Model Markup Language (PMML) which encodes the prediction models in Extensible Markup Language (XML)-based documents.

Now we have learnt the potential of big data in healthcare. In the next section, we undertake a case study to demonstrate how big data can contribute to healthcare in a real-life scenario.

7.2   Predicting days in hospital (DIH) using health insurance claims: a case study

Healthcare administrators worldwide are striving to lower the cost of care whilst improving the quality of care given. Hospitalization is the largest component of health expenditure. Therefore, earlier identification of those at higher risk of being hospitalized would help healthcare administrators and health insurers to develop better plans and strategies. In this section, we follow the work of Xie et al. (2015) and introduce a method to predict the number of hospitalization days in a population, using large-scale health insurance claim data.

7.2.1 Data manipulation and aggregation

This case study considered healthcare records generated when hospitals send claims to a health insurance company to receive reimbursement for their services. Data were available for a total of 3 years, with the periods from the initial 2 years (01/01/2010 to 31/12/2011) serving as a 2-year observation period and a 1 year prediction period from 1/1/2012 to 31/12/2012. Information from the 2-year observation period was used to predict outcomes in the third year.

The dataset included the hospital claims data for 242,075 individuals from a private health insurer named Hospitals Contribution Fund of Australia (HCF), which is one of Australia's largest combined registered private health fund and life insurance organizations. The dataset included hospital admission administrative records and hospital procedure claims, as well as enrolment information during the period an individual or his/her family was covered by the insurance policy. The data also contained basic demographic information of customers, such as age and gender (the dataset was received after pseudonymization to prevent the possibility of identifying individuals).

Before building a predictive model, the raw data were recoded and organized into a structure that allowed for efficient computing.

7.2.1.1 Three levels of information

The observed data have three levels of information: customer level, hospital admission level and hospital procedure claim level. These levels can be described as follows:

- *Customer level* contains customer demographics (e.g. age and gender) and information about the health insurance products they purchased. Each customer had at least one entry of such information. Customers who updated their personal information during the study period may have multiple records, while in such cases only the first record was taken, that is, the nearest record to the beginning of the study period.
- *Hospital admission level* includes hospital admission claims which provide detailed information about the customer and the healthcare provider. Key data elements contained in hospital admission claims include primary diagnosis, primary procedure, secondary diagnosis and the provider. Hospital admissions can be categorized broadly as inpatient or outpatient. Since HCF does not cover outpatient services, this part of data was not available. Inpatient claims are generated as a result of an admission to a facility involving same-day discharge or an overnight stay. Because of the duration of inpatient admissions, there can be more than one admission claim record; these records were aggregated to one unique entity for each admission. When calculating days in hospital (DIH), both same-day

admissions and 1 day overnight admissions were counted as 1 day. A customer may have multiple admissions within each year, and thus, have multiple entities at this level.
- *Hospital procedure claim level* contains hospital services delivered during a hospital admission. It includes service item information such as item type (e.g. same-day accommodation, overnight accommodation, prosthesis, theatre, etc.). It also contains information on the cost of each procedure. Each procedure claim is related to a hospital admission record, and a hospital admission record can be associated with multiple procedure claims.

Note that only the services covered by HCF were included in this database – in Australia, Medicare (Australia's social health insurer) concurrently covers most nonhospital services, and these data were not available.

7.2.1.2 Definition of key clinical data elements

There were five data elements that contained medical information involving different coding schemes:

- *Primary diagnosis code*: Each admission was associated with a diagnosis or condition which was considered chiefly responsible for the hospital admission. It was coded using ICD-10-AM (International Statistical Classification of Diseases and Related Health Problems, Tenth Revision, Australian Modification) codes (National Centre for Classification in Health, 2005). This is abbreviated as ICD-10 hereafter.
- *Primary procedure code*: Admissions also included a code for the procedure that consumed the largest amount of hospital resources. The corresponding coding scheme was developed by the Australian Classification of Health Interventions (ACHI) (National Centre for Classification in Health, 2010).
- *Diagnosis-related group*: Code of the Australian Refined Diagnosis-Related Group (AR-DRG) was associated with hospital admissions (Australian Institute of Health and Welfare, 2013) to classify acute inpatients. It was based on the codes allocated to the diagnoses and procedures for each episode of care.
- *Illness code*: Medicare Benefits Schedule (MBS) codes were used, which list the Medicare services subsidized by the Australian government, including a wide range of consultations, procedures and tests, as well as a schedule fee for each of these items (Department of Health, 2014). If no MBS code was found, other codes like ICD-10 and AR-DRG were considered instead. Various treatment types, illness groups and specialty descriptions were linked to the corresponding illness codes.
- *Item code*: Procedure items are of various types, which describe the nature of the service delivered or procedure performed. Each procedure or service has an item code expressed in a particular format, depending on the type of the item. The implication of particular characters may vary from one item to another. For example, for theatre, as a general rule, primary patient classification and hospital type indicate the theatre fee band, whereas for prostheses the accommodation type and primary patient classification imply the supplier.

Secondary diagnosis codes were also associated with admissions, and were only used to compute the Charlson Index which will be described in Section 7.2.1.4. Since only a small proportion of diagnosis codes could be used to generate the Charlson Index, secondary diagnosis codes were not introduced as major clinical elements. Instead, they are displayed in the final feature table, Table 7.2.

Table 7.2 Summary of the composition of the feature set. the four feature subset classes are: 1) demographic features; 2) medical features extracted from clinical information; 3) prior cost/dih features; and; 4) other miscellaneous features

Source variable	Level[1]	Type[2]	Computation[3]	Expansion[4]	Aggregation[5]	Number of features	Feature subsets
Year of the occurrence of hospital admission (2010, 2011 or 2012)	M	N	N/A	N/A	N/A	1	1
Age	M	N	Age 10; Age 60	B (Age 10)	N/A	16	1
Status of health policy held by customer	M	C	N/A	B	N/A	5	1
Type of membership (e.g. new customer, previous dependent in health policy, …)	M	C	N/A	B	N/A	6	1
Post code of address	M	C	N/A	N/A	N/A	1	1
Health insurance product	M	C	N/A	B	N/A	10	1
Dependent relationship in policy held by customer (e.g. policy holder, spouse, …)	M	C	N/A	B	N/A	14	1
Scale (family or single)	M	C	N/A	N/A	N/A	1	1
Gender	M	C	N/A	B	N/A	5	1
State code of address	M	C	N/A	B	N/A	10	1
Title code (e.g. Mr, Ms, …)	M	C	N/A	B	N/A	30	1

Attribute			Count	N/A	Sum		
Number of admissions for each customer in each year	A	N	N/A	N/A		1	2
Contract status (indicates if the procedure claim was paid under an HCF arrangement)	A	C	N/A	B	Ab	3	2
DAYS2PREV: days to previous admission	A	N	N/A	N/A	An	16	2
Diagnosis-related group	A	C	DRG_BASIC; DRG_CC; DRG_MDC; DRG_TEXT; DRG_subMDC	B (DRG_BASIC; DRG_CC; DRG_MDC; DRG_subMDC)	Ab (DRG_BASIC; DRG_CC; DRG_MDC; DRG_subMDC)	114	2
Type of hospital (public or private)	A	C	N/A	B	Ab	3	2
Primary diagnosis code	A	C	N/A	B; D	Ab; Ac	367	2
Primary procedure code	A	C	N/A	D	Ab; Ac	54	2
Actual days of the stay	A	N	N/A	N/A	An	18	3
Major disease category	A	C	N/A	D	Ab	25	2
Month of the hospital admission	A	C	N/A	B	Ab	15	2
Type of agreement between HCF and hospital	A	C	N/A	B	Ab	6	2

(Continued)

Table 7.2 (Continued)

Source variable	Level[1]	Type[2]	Computation[3]	Expansion[4]	Aggregation[5]	Number of features	Feature subsets
Type of care provider	A	C	N/A	N/A	Ac	4	2
Specific categorization of hospitals (depicting hospital sizes)	A	C	N/A	B	Ab	11	2
Secondary diagnosis and procedure	A	C	Charlson index; Number of secondary diagnoses; Number of secondary procedures	N/A	An (Charlson index)	18	2
Description of specialty group	A	C	N/A	B	Ab	57	2
Type of stay status in hospital (e.g. same-day, overnight accommodation, . . .)	A	C	N/A	B	Ab	7	2
Type of treatment (e.g. surgical, obstetric, . . .)	A	C	N/A	B	Ab	6	2
Amount charged for each procedure	P	N	N/A	N/A	An	17	3
Benefit paid at basic rate for each procedure (how much HCF paid)	P	N	N/A	N/A	Sum	1	3
Type of contract between care providers and HCF	P	C	N/A	B	Ab	19	2

Benefit paid at supplementary rate for each procedure	P	N	N/A	N/A	Sum	1	3
Type of hospital service item (e.g. accommodation, theatre, prosthesis, . . .)	P	C	N/A	B	Ab	17	2
Item code (code for procedure items)	P	C	IC_ACC_TYPE; IC_PAT_CLASS; IC_THEATRE; IC_PROSTHESIS	B (IC_ACC_TYPE; IC_PAT_CLASS)	Ab (IC_ACC_TYPE; IC_PAT_CLASS)	29	2
Count of hospital procedure claims for each customer in each year	P	N	Count	N/A	Sum	1	2
Patient classification code used to determine hospital accommodation benefits	P	C	N/A	N/A	Sum	1	2
Multiple source elements used	A&C	N; C	Admission moment features; Number of admissions in each year times age; Day of the admission (e.g. Monday, Tuesday, . . .)	N/A	Sum	5	4

(1) Information level of source variable: customer (M), admission (A), procedure (P); (2) Type of source variable: numberic (N), categorical (C); (3) Computation methods refer to Section 7.2.1.4; (4) Expansion method: binary expansion (B), dominant expansion (D), not applicable (N/A); (5) Aggregation method: aggregation of binary features (Ab), in which descriptive statistical operator sum was employed; aggregation of numeric features (An), in which descriptive statistical operator employed included count of elements, sum, mean, standard deviation, median, maximum and minimun, median and the most recent element; aggregation of categorical features (Ac), in which descriptive statistical operator used were sum, count of unique elements, the most frequent element, and the most recent element; not applicable (N/A); if not one of the above, features were simply summed up when being aggregated.

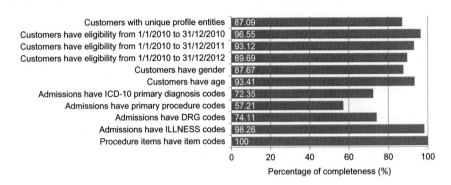

Figure 7.2 Data Completeness measured using the method described in Section 7.2.1.3. The numbers indicate percentage of completeness.

7.2.1.3   Data completeness

Data completeness refers to the indication of whether all the data necessary to answer the specified research question(s) are available or not. The completeness was estimated by measuring indicators shown in Figure 7.2. Out of 242,075 customers, 233,716 (96.55%) had eligibility for at least 1 year from 1/1/2010 to 31/12/2010, 225,421 (93.12%) had eligibility for at least two complete years from 1/1/2010 to 31/12/2011 and 217,111 (89.69%) customers had eligibility extending over the entire 3-year period. Naturally, the customer base was not constant as customers entered and left the fund from time to time. From a modelling perspective, an ideal situation would be that all subjects stay in the fund for the whole 3 years under analysis. However, since the customer cohort changes, predictions may also be made for some customers who would no longer be with the fund in the following year. Therefore, it is not necessary for each customer to have a complete 3-year full coverage. In the study, the customers who enrolled no later than 1/1/2010 were considered. Moreover, as observed from Figure 7.2, 87.09% of customers had unique profile entities without duplicates. 87.67% of customers had gender information and 93.41% had age information. Regarding the five medical data elements, 72.35% of admissions had clear ICD-10 primary diagnosis codes, 74.11% had DRG codes, 98.26% had illness codes and 57.21% had primary procedure codes. Nearly 100% of hospital procedure claims were assigned item codes. The completeness of DRG codes, primary diagnosis codes and primary procedure codes was not perfect because public hospital visits and procedures claimed through Medicare were not available to the HCF, as HCF and Medicare do not share these data.

7.2.1.4   Feature extraction

The source variables contained various types of data, such as dates, numeric and text. Therefore, at the beginning, source variables must be preprocessed before they can be used for modelling. Numeric variables were kept in their numeric format, whereas categorical variables were handled by means of quantification. After preprocessing, source variables were manipulated at their original information levels to generate additional features through processes of computation and expansion.

•   *Computation* processes were carried out to generate additional features which needed further calculation. In addition to its numeric format, age was categorized into 10-year bins, and a categorical feature named 'Age10' was generated. A binary indicator of age greater than 60 years or not was also taken into account. In addition to the numerical representations of the primary AR-DRG codes (DRG_TEXT and DRG_BASIC), they were decoded into additional features, for example, major disease category (DRG_MDC), major disease subcategories (DRG_subMDC) and comorbidity or complication (DRG_CC) (Australian Institute of Health and Welfare, 2013). In addition to the ICD-10-coded primary diagnosis, the list of ICD-10 coded secondary diagnoses were available for each hospital admission. From the secondary diagnoses, comorbidity scores were computed, utilizing the respective look-up tables, that give weights to certain ICD-10 diagnoses according to the original and the updated Charlson Index (Quan et al., 2011; Sundararajan et al., 2004), respectively.

    Two 'moment' features were computed to weight the number of admissions in a given month by the month's count number, that is, 1 to 12 in a linear (first moment) or quadratic (second moment) form, respectively. These features were employed to recognize the fact that admissions that occurred at the end of the year are likely to be followed by admissions in the following year. For hospital accommodation items, features of accommodation type (IC_ACC_TYPE) and patient classification (IC_PAT_CLASS) were extracted from accommodation item codes. Features of theatre fee band (IC_THEATRE) and supplier of prosthesis (IC_PROSTHESIS) were also extracted.

•   *Expansion* was applied to categorical variables. For those without too many unique categories, such as 'Age10' and gender, a binary expansion was performed to generate additional binary grouping features. Each grouping feature was a binary indicator with '1' indicating that entities fell into this category and '0' indicating that they did not.

    For those categorical variables with a large number of unique categories, additional features were developed to reduce the number of categories to a moderate level, using an external hierarchy grouping scheme. For instance, 287 groups of the ICD-10 primary diagnosis codes were considered rather than the 6292 individual entries (World Health Organization, 2010). Both the source variables containing specific codes and newly developed hierarchy categorical variables consisting of the reduced categories were kept. A binary expansion was applied afterwards to the hierarchy of categorical variables and 287 binary ICD-10 primary diagnosis grouping variables were added to the feature matrix. A dominant expansion of categorical variables was also considered. For instance, the 50 most frequently occurring ICD-10 primary diagnosis codes were chosen to be expanded to 50 binary grouping features. These 50 specific ICD-10 codes were selected by setting a criteria that each selected code should have at least 100 hospital admissions associated with it.

•   *Aggregation*: As described in Section 7.2.1.1, three levels of information exist: customer level, hospital admission level and hospital procedure claim level. Since the number of DIH was predicted at the customer level, further investigation should be performed at the other two levels. To aggregate numeric features extracted from the admission and the procedure claim levels with those from the customer level, descriptive statistics were employed, including sum, mean, median, mode, standard deviation, maximum, minimum, range, etc. Binary grouping features generated at the expansion stage were summed when aggregating with the customer-level features. Note that the aggregation of information at the procedure level went directly up to the customer level without going through the admission level in the middle.

7.2.1.5 Summary of the extracted features

A total of 915 features were present in the feature matrix. Table 7.2 summarizes all the features considered in this study. These features were categorized into four subsets based on the type of source variables from which they were extracted, which are in turn (i) demographic features, including personal information and enrolment information, (ii) medical features, containing clinical information extracted from diagnosis or procedure codes, (iii) prior cost/DIH features and (iv) other miscellaneous features.

Figure 7.3 demonstrates how the prediction model was built using information from the 2-year observation period to predict outcomes in the third year.

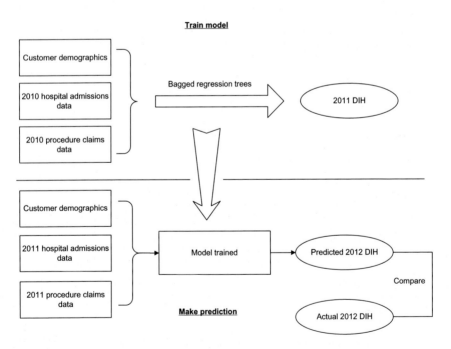

Figure 7.3 Customer demographics, admission and procedure claim data for year 2010, along with 2011 DIH were used to train the model. Later, at the prediction stage, customer demographics, hospital admission and procedure claim data for year 2011 were used to predict the number of DIH in 2012.

7.2.2 Predictive methods

7.2.2.1 Baseline models

To carry out meaningful comparisons, two methods were used as baselines so that comparisons could be made between the prediction models developed. The first baseline model predicted the same constant number of days for all customers. The number of days selected for this constant was chosen to optimize one of the performance measures — root-mean-square-error (RMSE) which will be described later in Section 7.2.3.1. In the second baseline model, the DIH of the second year (2011) of the observation period was used as the forecast for the DIH in the prediction year (2012).

7.2.2.2 Bagged trees

A predictive model was built using bagged regression trees, which is computationally efficient to train on large datasets (Breiman, 2001; Breiman & Cutler, 2014). Every tree in the ensemble is grown on an independently drawn bootstrap replica of the data. Observations not included in this replica are 'out-of-bag' samples for this tree. To compute a prediction from an ensemble of trees for unseen data, the average of predictions is taken from each individual tree in the ensemble. The out-of-bag observations are used as a validation dataset to estimate the prediction error, which helps to avoid severe overfitting. Here a function named 'treebagger' in the statistics toolbox of MATLAB R2013b (MathWorks, Natick, MA, USA) was used to implement the algorithm (The Mathworks, 2014). After tuning parameters, the number of trees used in the bagging ensemble was 100, and the minimum number of observations (customers) per tree leaf was 5. Other parameters, if not specified, were default MATLAB settings (The Mathworks, 2014).

7.2.3 Performance measures

7.2.3.1 Performance metrics

Table 7.3 gives an overview of the performance indicators used in this study. The main performance indicator is referred to as the RMSE, and is the root-mean-square of the difference between the logarithm of the estimated DIH and the logarithm of the true number of days (Brierley, Vogel, & Axelrod, 2013). The logarithm (offset by $+1$ to avoid a logarithm of zero) was used to reduce the importance assigned to those with many hospital days. In order to capture different aspects of the goodness of predictability, additional performance measures were also reported. Table 7.3 lists the measures and their equations. In addition to RMSE, the Spearman rank correlation coefficient (ρ) was also calculated between the predicted and actual number of DIH.

Customers can be categorized into two groups, those without hospital days (0 hospital days) and those with hospital days (at least 1 hospital day) per year. Therefore, binary classification analysis is applicable to the result. By setting a threshold for the predicted number of hospital days, statistics including accuracy (Acc.), specificity (Spec.), sensitivity (Sens.), Cohen's kappa (κ) and area under receiver operating characteristic curve (AUC) metrics were calculated. The optimal results were obtained

Table 7.3 Performance measures

Measure	Equation
RMSE	$\sqrt{\frac{1}{N}\sum_{i=1}^{N}\left[\ln(p_i+1)-\ln(a_i+1)\right]^2}$
ρ	$1-\frac{6\sum(\ln(p_i+1)-\ln(a_i+1))^2}{N(N^2-1)}$
AUC	$\int_0^1 R_{tp}(T)R'_{jp}(T)dT$
Spec.	$\frac{N_{tn}}{N_{tn}+N_{fp}}$
Sens.	$\frac{N_{tp}}{N_{tp}+N_{fn}}$
Acc.	$\frac{N_{tp}+N_{tn}}{N}$
κ	$\frac{P_o-P_e}{1-P_e}$
R^2	$1-\frac{\sum(a_i-p_i)^2}{\sum(a_i-m)^2}$
$\|R\|$	$1-\frac{\sum|a_i-p_i|}{\sum|a_i-m_{median}|}$

N is the total number of customers in the population; p_i is the predicted number of DIH for the ith person; a_i is actual DIH, $i \in [1, N]$; m is the average DIH for population; m_{median} is the median value of DIH for population; $R_{tp}(T)$ and $R_{fp}(T)$ are the true positive rate (sensitivity) and the false positive rate (equals to $1-$ specificity) for a given threshold T in a binary classification model; N_{tp} is the number of hospitalized patients who were correctly predicted as having ≥ 1 DIH; N_{tn} is the number of subjects who were correctly predicted as having 0 DIH; N_{fp} is the number of subjects who were predicted as having ≥ 1 DIH, but actually had 0 DIH; N_{fn} is the number of subjects who were predicted as having 0 DIH, but actually had ≥ 1 DIH. P_o is the ratio of the probability of observed agreement between observation and prediction $\left(\frac{N_{tp}+N_{fn}}{N}\right)$; P_e is the probability of random agreement $\left(\frac{N_{tp}+N_{fn}}{N} \times \frac{N_{tp}+N_{fp}}{N}\right) + \left(\frac{N_{fp}+N_{tn}}{N} \times \frac{N_{fn}+N_{tn}}{N}\right)$

with a threshold of 0.3 days applied to the continuous output estimate from the bagged tree regression model. When the predicted value was smaller than 0.3 days, it was considered as a prediction of no hospital days and vice versa. The choice of threshold was empirically based on the fact that using 0.3 days as threshold gave the best κ.

The coefficient of determination (R^2) (Cumming, Knutson, Cameron, & Derrick, 2002) was also calculated to show the ability of this model to explain the variation. Bertsimas et al. (2008) proposed a new measure of $|R|$ adjusted from R^2. R^2 measures the ratio of the improvement of predictability (which is approximated by the sum of the squares of the residuals) of a regression line compared with a constant prediction (the mean of actual values), while $|R|$ measures the reduction in the sum of absolute values of the residuals compared with another constant prediction (the median of actual values).

The importance of features was also evaluated using the criterion ('OOBPermutedVarDeltaError') in the MATLAB 'treebagger' function. It gives a numeric array containing a measure of importance for each feature. For any variable, the measure is the increase in prediction error if the values of that variable are permuted across the out-of-bag observations. This measure is computed for every tree, then averaged over the entire ensemble and divided by the standard deviation over the entire ensemble.

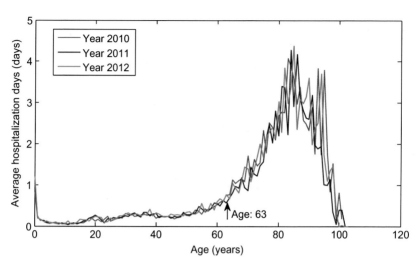

Figure 7.4 Average days in hospital per person by age for each of the 3 years of HCF data.

7.2.3.2   Subpopulations

The performance was measured on four different scales of population. Group 1 was set equal to the entire population. Group 2 included customers born in or after the year 1948, while Group 3 consisted of customers born before the year 1948. These subgroupings were chosen since the average number of DIH increased substantially between the ages of 61 and 65 years, as Figure 7.4 indicates, and the median age of 63 years in year 2011 was taken, corresponding to a birth year 1948 in this dataset. In the whole population, 85% of customers were born in or after 1948 and 15% of customers were born before 1948. In addition, the model was evaluated on a subpopulation (Group 4) in which customers had at least 1 day (1+ days) in hospital in the year before the prediction year (2012). It should be noted that the 1 + days group was the only subpopulation in which the cohort of customers was not consistent over the training and prediction sets. In this group, the training process was performed using the data of those customers who had at least 1 day in hospital in 2010, whereas the prediction was performed on customers who had at least 1 day in 2011.

7.2.4   Results
7.2.4.1   Performance on different subpopulations

Table 7.4 summarizes the performance of the proposed methods evaluated on the four subpopulations, as discussed in the previous section. The results from the training and prediction datasets, as well as those of the two baseline methods are presented.

Table 7.4 Performance metrics of the proposed method, evaluated on different populations

Model/dataset	RMSE	ρ	AUC	Spec. (%)	Sens. (%)	Acc. (%)	κ	R^2	|R|
All subjects									
Train[a]	0.281	0.458	0.958	94.47	72.97	92.56	0.595	0.287	0.087
Predict	0.381	0.344	0.843	91.81	43.17	87.50	0.311	0.150	−0.092
Baseline 1	0.428	N/A	N/A	N/A	N/A	N/A	N/A	−0.007	−0.183
Baseline 2	0.490	0.272	0.634	93.46	32.94	88.10	0.264	−0.205	−0.462
Subjects born in or after 1948									
Train	0.237	0.419	0.964	97.32	62.12	94.81	0.602	0.296	0.009
Predict	0.305	0.288	0.820	95.15	26.28	90.28	0.225	0.109	−0.186
Baseline 1	0.330	N/A	N/A	N/A	N/A	N/A	N/A	−0.003	−0.361
Baseline 2	0.399	0.211	0.604	94.34	26.29	89.53	0.206	−0.365	−0.567
Subjects born before 1948									
Train	0.477	0.644	0.949	72.82	97.45	77.86	0.508	0.270	0.160
Predict	0.701	0.449	0.814	65.45	81.64	68.77	0.330	0.130	−0.023
Baseline 1	0.787	N/A	N/A	N/A	N/A	N/A	N/A	−0.023	−0.128
Baseline 2	0.879	0.361	0.677	86.66	48.00	78.74	0.347	−0.166	−0.372
1 + days group (all ages)									
Train	0.440	0.776	0.965	78.40	98.93	85.01	0.693	0.382	0.276
Predict	0.780	0.369	0.703	47.79	78.94	58.02	0.219	0.191	0.013
Baseline 1	0.890	N/A	N/A	N/A	N/A	N/A	N/A	−0.032	−0.080
Baseline 2	1.238	0.208	N/A	N/A	N/A	N/A	N/A	−0.351	−0.904

Shown are the results for testing with the training data on the model after training (with the same data) and results when validating the trained model with the prediction dataset.
Performances of baseline models 1 and 2 are also displayed.
[a]The 'Train' model was tested with the same training data that was used to train it to give a measure of fit, whereas the other three models are all tested using data from the third year.

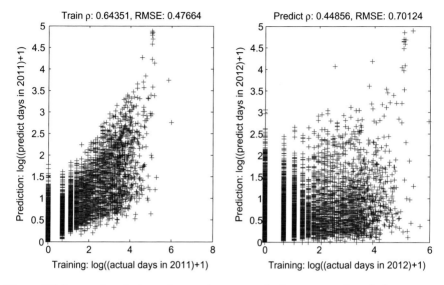

Figure 7.5 Scatter-plots for bagged regression tree results for customers born before year 1948 (those aged 63 years or older when the model was trained in 2011).

Figure 7.5 exhibits the scatter-plots of the regression results for the group born before the year 1948. The subplot on the left is for training, while the other is for prediction.

7.2.4.2 Evaluation of features

The models' predictive capability was investigated using the subsets of features discussed in Section 7.2.1.5 and then compared to the prediction results when all the features were taken into account. These results are displayed in Table 7.5 for algorithms using demographic, medical and past cost and DIH features separately. Since the miscellaneous features comprise a small group of heterogeneous features only, they were not analysed.

The top 200 features for the four subpopulations mentioned in Section 7.2.3.2 were also examined. Table 7.6 lists some interesting top features extracted from ICD-10 primary diagnosis code derived on the 1 + days group. Age was used as a reference feature for comparison.

Figure 7.6 shows how the top 200 features are distributed among the four feature subsets. Figure 7.6(a)−(d) display the distribution of the top 200 features across the entire population, subjects born in or after 1948, subjects born before 1948, and 1 + days group, respectively. Figure 7.6(e) shows the proportion of all features among the four subsets.

Table 7.5 Performance metrics of predictions using feature category subsets only

Feature subset		RMSE	ρ	AUC	Spec. (%)	Sens. (%)	Acc. (%)	κ	R²	|R|
All subjects	Demographic	0.393	0.337	0.838	90.07	45.80	86.15	0.295	0.030	−0.142
	Medical	0.396	0.273	0.635	95.61	28.00	89.60	0.268	0.136	−0.135
	Cost and DIH	0.400	0.272	0.635	95.44	27.18	89.39	0.256	0.093	−0.151
	All	0.381	0.344	0.843	91.81	43.17	87.50	0.311	0.150	−0.092
Subjects born in or after 1948	Demographic	0.315	0.284	0.816	94.23	26.85	89.46	0.208	0.019	−0.231
	Medical	0.312	0.210	0.604	97.47	17.25	91.79	0.191	0.100	−0.235
	Cost and DIH	0.315	0.209	0.604	97.34	15.76	91.57	0.170	0.088	−0.247
	All	0.305	0.288	0.820	95.15	26.28	90.28	0.225	0.109	−0.186
Subjects born before 1948	Demographic	0.725	0.415	0.793	58.20	89.58	64.64	0.306	0.011	−0.068
	Medical	0.723	0.365	0.679	86.99	47.27	78.85	0.346	0.126	−0.045
	Cost and DIH	0.729	0.362	0.678	86.95	47.15	78.79	0.344	0.069	−0.061
	All	0.701	0.449	0.814	65.45	81.64	68.77	0.330	0.130	−0.022
1 + days group (all ages)	Demographic	0.839	0.283	0.653	36.56	81.56	51.34	0.141	0.013	−0.066
	Medical	0.801	0.319	0.676	38.24	81.99	52.61	0.159	0.183	−0.001
	Cost and DIH	0.825	0.248	0.632	39.69	75.85	51.57	0.124	0.110	−0.032
	All	0.780	0.369	0.703	47.79	78.94	58.02	0.219	0.191	0.013

The three feature subsets tested are: demographic features, medical features and prior cost/DIH features. Subset of miscellaneous features were not used.

Table 7.6 An example of interesting ICD-10 primary diagnosis features

Feature	Measure of importance
ICD10_PREGN_E_02 (Oedema, proteinuria and hypertensive disorders in pregnancy, childbirth and the puerperium)	1.188
ICD10_PRINC_DIAG_N979 (Female infertility, unspecified)	0.883
ICD10_PRINC_DIAG_Z312 (In vitro fertilization)	0.673
ICD10_PRINC_DIAG_C61 (Malignant neoplasm of prostate)	0.402

Age was used as a reference feature with an importance measure of 0.702. The definition of this importance measure was described in Section 7.2.3.1.

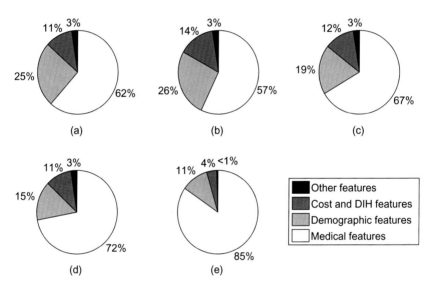

Figure 7.6 Distribution of the top 200 features among the four feature subsets: (a) in the whole population; (b) in subjects born in or after 1948; (c) in subjects born before 1948; (d) in the 1+ days group; (e) shows the percentage of the four subsets with respect to the full feature set of 915 features.

7.2.5 Discussion and conclusion

An approach to predicting future DIH was proposed using features extracted from customer demographics, past hospital admission and hospital procedure claim data. The predictive model was trained using data from an observation period of 2 years and then evaluated based on the data from the third year. In summary, Table 7.4 indicates that: (a) the proposed method improved the prediction results over both of the baseline methods (as described in Section 7.2.2.1) by reducing the RMSE measure and increasing R^2 and $|R|$; (b) from the view of the coefficients of determination, R^2 and $|R|$, this method achieved the best performance on the 1 + days group, with an R^2 of 19.1% compared to 15.0% for all subjects, 10.9% for the young subpopulation and 13.0% for the group of customers born in or before 1948.

Table 7.4 also shows that the model achieved a high specificity (95.15%) and a relatively low sensitivity (26.28%) on the test set of customers born in or after 1948. There could be a variety of reasons behind this low sensitivity. One of the possible explanations is that younger customers are more likely to be hospitalized for acute conditions which are unpredictable or unexpected; therefore, it may be inherently more difficult to make true positive predictions for younger people. In contrast, for the aged group, a moderate specificity (65.45%) and high sensitivity (81.64%) were obtained. The specificity dropped in comparison to the younger group and the whole population. A similar pattern was exhibited in the 1 + days group, with a lower specificity (48.05%) and a relatively high sensitivity (78.52%). The underlying reason may be that people with medical conditions in the previous years were well controlled and therefore are less likely to be hospitalized in the future. As a consequence, although the model predicted that the customer would be hospitalized, it tends not to happen.

Table 7.5 indicates that for the 1 + days group, when the three subsets of features were considered independently, the medical feature subset achieved the best RMSE, R^2, κ and ρ. For the other three groups, the difference in the performance of the three subsets of features was not as obvious as that of the 1 + days group. However, taking into account all the features always led to the best performance, implying the presence of the synergy among predictors from different subsets.

When evaluated on the 1 + days group, the importance measure for age was around 0.702 with a ranking of 22 out of 915. As shown in Table 7.6, the ICD-10 diagnosis code related to pregnancy with oedema, proteinuria and hypertensive disorders in pregnancy, childbirth and the puerperium, exhibited a surprisingly high importance of 1.188. ICD-10 diagnosis codes N979 (female infertility, unspecified) and Z312 (in vitro fertilization) resulted in importance measure values of 0.882 and 0.672, respectively, while a diagnosis code about malignant neoplasm of prostate (C61) was ranked at 68 out of 915 (importance measure value = 0.402). As mentioned in Section 7.2.1.4, an expansion of the most frequent ICD-10 codes was performed. Fifty specific codes were expanded to grouping variables, but only those three came in the top 200. However, plotting similar regression figures to Figure 7.4 for these ICD-10 diagnosis features and DIH revealed no strong linear correlations. A hypothesis for why these codes are highly ranked is that they are associated with rather homogeneous patient groups without many comorbidities, and therefore are more predictive.

Figure 7.6 conveys the following information: (i) although the prior cost and DIH features only comprised 4% of all features, they stably contributed 11% or more to the top 200 features for all the four customer groups, implying the strong predictive power of such features; (ii) from Figure 7.6(b)–(d), the percentage of demographic features gradually declines while the proportion of medical features increases. This is in accord with the fact that the density of medical information increases among the three subpopulations. The 1 + days group had the densest medical information; therefore medical features greatly improved the prediction for this group. For subpopulations where medical information was not plentiful, such as the subjects born in or after 1948, only 57% of the top features were medical features and prediction relied heavily on demographic features, past cost and DIH features.

To validate the performance of the prediction of DIH, one needs to pay attention to the completeness and quality of insurance claims data. Since the purpose of insurance claims is generally to carry out the reimbursement, the data of procedure items and the associated costs are complete in almost all cases. Nonetheless, other relevant information, such as diagnosis codes, is not always available to the analyst. Furthermore, coding errors are not rare in the insurance claims data. For instance, for reimbursement purposes a physician may code a diagnosis in the absence of the confirmation of medical test results, which is often the case for emergency treatments (Duncan, 2012). In practice, it is advised that one should examine the completeness of the insurance claims data before fitting the model, while the quality of data should be reported along with the results of analyses.

The predictive model considered in this chapter can be applied in various fields. For example, insurance companies can manage the risk of insurance by undertaking specific interventions with respect to high-risk customers. If those individuals who tend to be hospitalized frequently can be identified by the predictive model prior to their episodes of high utilization of medical resources, they can be engaged in healthcare management programs which not only improve the quality of their lives but also moderate the associated costs.

Acknowledgement

We would like to thank the Hospitals Contribution Fund of Australia for providing the data and financial support. We are grateful to the Austrian Institute of Technology (AIT) for their advice and technical support which have improved the quality of this chapter.

References

Akay, M. F. (March 2009). Support vector machines combined with feature selection for breast cancer diagnosis. *Expert Systems with Applications, 36*(2), 3240−3247 . [Online]. Available <http://www.sciencedirect.com/science/article/pii/S0957417408000912>
Amarasingham, R., Patel, P. C., Toto, K., Nelson, L. L., Swanson, T. S., Moore, B. J., et al. (Dec. 2013). Allocating scarce resources in real-time to reduce heart failure readmissions: A prospective, controlled study. *BMJ Quality & Safety, 22*(12), 998−1005 . [Online]. Available <http://qualitysafety.bmj.com/content/22/12/998.short>
Australian refined diagnosis-related groups (AR-DRG) data cubes. (2013). <http://www. aihw.gov.au/hospitals-data/ar-drg-data-cubes/> Australian Institute of Health and Welfare, Australian Government.
Bellazzi, R., & Zupan, B. (2008). *Predictive data mining in clinical medicine: Current issues and guidelines* (pp. 81−97).
Bertsimas, D., Bjarnadottir, M. V., Kane, M. A., Kryder, J. C., Pandey, R., Vempala, S., et al. (2008). Algorithmic prediction of health-care costs. *Operations Research, 56*, 1382−1392.
Breiman, L. (2001). Random forests. *Machine Learning, 45*(1), 5−32.
Breiman, L., & Cutler, A. (2014). *Random forests.* <http://www.stat.berkeley.edu/~breiman/ RandomForests/>.

Brierley, P., Vogel, D., & Axelrod, R. (2013). *Heritage provider network health prize round 1 milestone prize: How we did it — Team 'Market Makers'.* <http://www.heritage-healthprize.com/c/hhp/leaderboard/milestone1>.

Caroli, A., Manini, S., Antiga, L., Passera, K., Ene-Iordache, B., Rota, S., et al. (2013). Validation of a patient-specific hemodynamic computational model for surgical planning of vascular access in hemodialysis patients. *Kidney International, 84*(6), 1237–1245. [Online]. Available <http://www.ncbi.nlm.nih.gov/pubmed/23715122>.

Cottle, M., & Hoover, W. (2013). *Transforming health care through big data.* New York: Institute for Health Technology Transformation. Tech. Rep. [Online]. Available <http://c4fd63cb482ce6861463-bc6183f1c18e748a49b87a25911a0555.r93.cf2.rackcdn.com/iHT2_BigData_2013.pdf>.

Cumming, R. B., Knutson, D., Cameron, B. A., & Derrick, B. (2002). *A comparative analysis of claims-based methods of health risk assessment for commercial populations.* Chicago: Society of Actuaries.

De Choudrhury, M., Gamon, M., Counts, S., & Horvitz, E. (2013). Predicting depression via social media. *Seventh international AAAI conference on weblogs and social media* (Vol. 2, pp. 128–137).

Duncan, I. (2012). Mining health claims data for assessing patient risk. In D. E. Holmes, & L. C. Jain (Eds.), *Data mining: Foundations and intelligent paradigms, ser. Intelligent Systems Reference Library* (Vol. 25, pp. 29–62). Berlin Heidelberg: Springer.

Ginsberg, J., Mohebbi, M. H., Patel, R. S., Brammer, L., Smolinski, M. S., & Brilliant, L. (2009). Detecting inuenza epidemics using search engine query data. *Nature, 457*(7232), 1012–1014.

Herland, M., Khoshgoftaar, T. M., & Wald, R. (June 2014). A review of data mining using big data in health informatics. *Journal of Big Data, 1*(1), 2. [Online]. Available <http://www.journalofbigdata.com/content/1/1/2>.

Hermon, R., & Williams, P. (2014). Big data in healthcare: What is it used for? In *Proceedings of the third Australian eHealth informatics and security conference.* [Online]. Available <http://ro.ecu.edu.au/aeis/22>.

International statistical classification of diseases and related health problems, tenth revision. (2010). <http://apps.who.int/classifications/icd10/browse/2010/en> World Health Organization.

Klompas, M., Eggleston, E., McVetta, J., Lazarus, R., Li, L., & Platt, R. (April 2013). Automated detection and classification of type 1 versus type 2 diabetes using electronic health record data. *Diabetes Care, 36*(4), 914–921 . [Online]. Available <http://care.diabetesjournals.org/content/36/4/914.long>.

Klompas, M., McVetta, J., Lazarus, R., Eggleston, E., Haney, G., Kruskal, B. A., et al. (June 2012). Integrating clinical practice and public health surveillance using electronic medical record systems. *American Journal of Preventive Medicine, 42*(6 Suppl. 2), S154–S162. [Online]. Available <http://www.sciencedirect.com/science/article/pii/S0749379712002498>

Medicare benefits schedule (MBS). (2014). Department of Health, Australian Government. <http://www.mbsonline.gov.au/>.

Peng, R. D., Leek, J., & Caffo, B. (2015). <https://www.coursera.org/course/repdata> *Reproducible research.* Coursera.

Pentland, A., Reid, T., & Heibeck, T. (2013). Big data and HealtH: Revolutionizing medicine and public health. *MIT consortium for Kerberos and Internet Trust, Tech. Rep.* [Online]. Available <http://kit.mit.edu/sites/default/files/documents/WISH_BigData_Report.pdf>.

Quan, H., Li, B., Couris, C. M., Fushimi, K., Graham, P., Hider, P., et al. (2011). Updating and validating the Charlson comorbidity index and score for risk adjustment in hospital discharge abstracts using data from 6 countries. *American Journal of Epidemiology, 173* (6), 676−682.

Smith, S. M., Vidaurre, D., Beckmann, C. F., Glasser, M. F., Jenkinson, M., Miller, K. L., et al. (2013). *Functional connectomics from resting-state fMRI* (pp. 666−682).

Sundararajan, V., Henderson, T., Perry, C., Muggivan, A., Quan, H., & Ghali, W. A. (2004). New ICD-10 version of the Charlson comorbidity index predicted in-hospital mortality. *Journal of Clinical Epidemiology, 57*(12), 1288−1294.

The Australian classification of health interventions, seventh edition, tabular list of interventions and alphabetic index of interventions. (2010). Australia: National Centre for Classification in Health, University of Sydney. <http://meteor.aihw.gov.au/content/index.phtml/itemId/391343/>.

The international statistical classification of diseases and related health problems, tenth revision, Australian modification (4th ed.). (2005). Australia: National Centre for Classification in Health, University of Sydney. <http://meteor.aihw.gov.au/content/index.phtml/itemId/325387/>.

The Mathworks. (2014). Treebagger. <http://www.mathworks.com.au/help/stats/treebagger.html/>.

Viceconti, M., Hunter, P., & Hose, D. (2015). Big data, big knowledge: Big data for personalised healthcare. *IEEE Journal of Biomedical and Health Informatics, 19*(4), 1209−1215.

Weber, G. M., Mandl, K. D., & Kohane, I. S. (June 2014). Finding the missing link for big biomedical data. *JAMA, 311*(24), 2479−2480 . [Online]. Available <http://jama.jamanetwork.com/article.aspx?articleID=1883026>.

Xie, Y., Schreier, G., Chang, D. C. W., Neubauer, S., Liu, Y., Redmond, S. J., et al. (2015). Predicting days in hospital using health insurance claims. *IEEE Journal of Biomedical and Health Informatics, 19*(4), 1224−1233.

Zhao, Y., Ash, A. S., Ellis, R. P., Ayanian, J. Z., Pope, G. C., Bowen, B., et al. (2005). Predicting pharmacy costs and other medical costs using diagnoses and drug claims. *Medical Care, 43*, 34−43.

Big data from mobile devices

8

In the era of big data, there is an enormous amount of information that is being collected, communicated or analysed by mobile devices. Every month, there are about 1.5 billion gigabytes of data transferred over mobile networks. Due to the boom of the Internet of Things, the amount of mobile devices that connect to the internet will rise to about 50 billion by 2020, which will greatly increase the volume, variety and velocity of data.

In the context of big data analysis, there are two major groups of mobile devices: wearable sensors and mobile phones. The former is designed with a primary focus on monitoring health conditions of individuals. Pantelopoulos and Bourbakis (2010) stated that over the past years, the design and development of wearable sensors for health monitoring have received much interest from the scientific community and industry. Because of the increasing costs of healthcare, technologies in miniature sensing devices, smart textiles, microelectronics and wireless communications have been progressing continuously, implying that wearable sensors will potentially transform the future of healthcare by enabling proactive personal health management and ubiquitous health monitoring of individuals. Furthermore, sensor-based systems can be formed, which comprise various types of small physiological sensors, transmission modules or processing capabilities to facilitate low-cost wearable unobtrusive solutions for continuous all-day and any-place health, mental and activity status monitoring.

On the other hand, mobile phones (sometimes referred to as smart phones) are becoming the central computer and communication device in our lives. As stated by Lane et al. (2010), smart phones are programmable and equipped with powerful embedded sensors, including accelerometer, gyroscope, GPS, compass, microphone and camera. These sensors jointly enable innovative applications in a wide range of areas. For example, in social networks mobile phone sensors can be used to classify automatically the events people are participating in, and share the participation using online social networks. In health monitoring, sensor-enabled mobile phones are able to observe data continuously (often with high velocity), providing important information about patients' health status. For instance, physical activities can be captured and related to personal health goals and consequently feedbacks to users are generated, such as encouraging more exercise. In environmental engineering, mobile phone sensors can be utilized to evaluate the environmental impact of an individual, monitoring how the actions of an individual influence his/her exposure and contribution to environmental issues such as greenhouse gas emissions. An important advantage of sensor-equipped mobile phones is that they are much more efficient than traditional cell phones in data recording and communication. On average, a smart phone generates about 48 times more mobile data traffic than a basic cell phone. In 2014, about 27% of global handsets were smartphones, but they accounted for 95% of global handset traffic. The predicted number of smart phones in 2015 is over 1.2 billion, and the growth is expected to continue. The sensor-equipped mobile phones are believed to revolutionize data collection and processing.

As stated before, data collected by mobile devices often exhibit high-volume, high-variety and high-velocity characteristics. Suitable methods need to be applied to extract useful information from these data. In this chapter, we discuss issues related to data collection and processing using mobile devices. In turn, we concentrate on the applications of wearable devices in health monitoring, and a case study in transportation where data collected from mobile devices facilitate the management of road networks.

8.1   Data from wearable devices for health monitoring

As noticed by Patel, Park, Bonato, Chan, and Rodgers (2012), because of the improvements in healthcare over the past few decades, residents of industrialized countries are now living longer but with multiple health conditions which are often complex. Although the survival from acute trauma has improved, it has led to an increase in the number of individuals with disabilities. It is therefore of critical importance to seek solutions to the question of how healthcare can be provided to a growing amount of individuals with complex medical conditions or disabilities, while the cost remains affordable. Patel et al. (2012) argued that such a problem can be solved partially by utilizing advanced techniques of wearable devices, since evidence has shown that recent developments in wearable sensor-based systems have led to exciting clinical applications, such as diagnostics and monitoring.

Patel et al. (2012) suggested the following framework of health monitoring using wearable devices:

1. Wearable sensors are deployed according to the clinical application of interest. Such sensors are capable of monitoring vital signs and movements of patients;
2. Data collected by wearable sensors are transferred to a processing centre;
3. Emergency situations (e.g., heart failure, fall) are detected by data analysis, and a warning message is sent to an emergency service centre to provide immediate assistance to patients.

The three steps above correspond to three key enabling technologies in health monitoring: sensor technology, communication technology and data analysis techniques. The first two technologies are in the field of information technology or electronic engineering and hence out of the scope of this book. The main focus of this section is on the analysis techniques that are suitable for data collected by wearable devices. As stated by Patel et al. (2012), for the purpose of health monitoring, massive data that are gathered using wearable devices have to be managed and processed to derive clinically relevant information. Statistical methods, such as signal processing, pattern recognition and data mining, enable remote health monitoring applications which would have been infeasible otherwise. It is now widely acknowledged that data processing and analysis techniques should be integrated in the design and development of wearable-device based health monitoring. The following sections aim to demonstrate some of these techniques.

8.1.1 Signal, sensor and data

Table 1 of Pantelopoulos and Bourbakis (2010) provides a list of sensors that can be integrated as a part of wearable health-monitoring system. Each of these sensors corresponds to a certain type of biosignal being monitored, while records of such a biosignal are stored in data format. According to Pantelopoulos and Bourbakis (2010), the following types of biosignals are of interest in health monitoring:

- Electrocardiogram (ECG): it is measured by chest electrodes to monitor electrical activity of heart, with signals recorded as continuous waveform showing the contraction and relaxation phases of cardiac cycles.
- Electroencephalogram (EEG): recorded by scalp-placed electrodes, measuring electrical spontaneous brain activity and other brain potentials.
- Electromyogram (EMG): measured by skin electrodes to examine electrical activity of skeletal muscles.
- Heart rate: recorded by skin electrodes.
- Heart sounds: measured by phonocardiograph.
- Blood pressure: monitored by arm cuff-based sensors, recorded as the force exerted by circulating blood on the walls of blood vessels.
- Blood glucose: recorded by strip-based meters as the amount of glucose in blood.
- Body/skin temperature: monitored by temperature probe or skin patch.
- Respiration rate: recorded by piezoelectric sensors.
- Oxygen saturation: measured by pulse oximeter, indicating the amount of oxygen contained in blood.
- Perspiration or skin conductivity: measured by galvanic skin response.
- Body movements: monitored by accelerometers and gyroscopes. These sensors are used to record measurements of acceleration forces in the 3-dimensional space as well as information about the position of a patient.

In the past decade, various methods have been developed for the analysis of these recorded signals, aiming to facilitate health monitoring by abnormality detection. To illustrate, we consider data recorded by the following two types of sensors: electrodes and accelerometers/gyroscopes. Section 8.1.2 has a focus on electrode-type sensors, providing a discussion about statistical methods that are suitable for ECG, EEG or EMG data. Section 8.1.3 briefly describes how a fall of an individual being monitored can be detected based on the data collected from accelerometers and gyroscopes.

8.1.2 ECG, EEG and EMG

Using electrodes, ECG, EEG or EMG data are observed in the form of time series, that is, values are recorded over time with a fixed frequency. Time series data are ubiquitous in our lives. Aghabozorgi, Shirkhorshidi, and Wah (2015) noted that real-world applications have found the chance to store data for a long time, as a result of the development of data storage and processors. Consequently, large volumes of time series data are routinely created in various fields, such as economics (GDP, inflation rates), finance (stock prices, exchange rates), environmental studies (greenhouse gas emissions, ambient temperatures), seismology (seismic waves due to earthquakes), etc. In the context of biomedical research, signals recorded by electrodes are typical examples of time series data. The figure below displays time series patterns of electrocardiograms (ECG), electroencephalograms (EEG) and electromyograms (EMG).

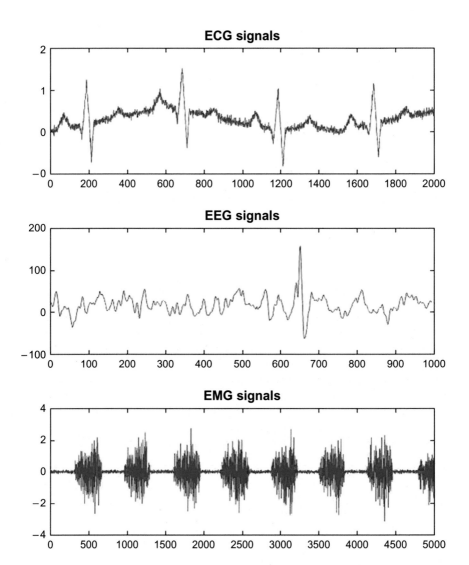

Groups of electrode signals have been extensively discussed in the literature. For ECG signals, there are groups associated with various types of cardiac arrhythmias including atrial premature contraction, premature ventricular contraction, supraventricular tachycardia, ventricular tachycardia and ventricular fibrillation, as well as a normal sinus rhythm (healthy people) group (Ge, Srinivasan, & Krishnan, 2002). For EEG signals, groups of electrode recordings include those of healthy volunteers, those of epileptic patients measured during seizure-free intervals, and those of epileptic patients measured during seizure activity (Maharaj, 2014). In the field of myoelectric prosthesis control, EMG signals collected from remnant muscles are used to define various types of hand motion (Mattioli, Lamounier, Cardoso, Soares, & Andrade, 2011). In health monitoring, discriminating between groups of electrode signals is central to the detection of particular illness with respect to those of healthy people. Such a discriminating task is known as classification, or supervised learning. The aim of classification is to assign an individual (i.e., a patient), whose grouping is unknown *a priori*, to one of several existing groups. The determined grouping of this individual indicates his/her health status. For example, the classification of EEG recordings of an epileptic patient is able to tell whether he/she is during seizure activity or not. It should be stressed that the analysis of electrode signals is usually automated when health monitoring is carried out remotely. As a result, severe adverse consequences may happen if electrode signals are incorrectly classified. Taking ECG as an example, it is very risky if a classification method cannot discriminate between signals of normal sinus rhythm and those of arrhythmias, as arrhythmias represent a serious threat to the patient recovering from acute myocardial infarction (Ge et al., 2002). Consequently, it is essential to consider classification methods that produce low rates of false negatives, that is, the result of classification incorrectly indicates no presence of a condition (e.g., seizure activity, acute myocardial infarction, etc.) but in reality it is present.

Aghabozorgi et al. (2015) claimed that over the past ten years a large amount of changes and developments have emerged in time series analysis, which have increased the size of time series datasets exponentially. Taking electrodes as an example, one hour of ECG records occupies one gigabyte. To cope with such a trend, methods for classifying time series data efficiently are required. In this section, feature-based classification methods are discussed due to their reliability and low computation cost.

Feature-based classification methods decide on the grouping of a time series based on some form of features extracted from data. These methods are usually straightforward to implement as most of time series features can be computed easily. The benefit of using feature-based methods is substantial when the length of time series is very large, since one may reduce the size of data by characterizing each time series using a small number of suitable features, rather than working directly on massive raw values. Another advantage is that feature-based methods are capable of classifying time series with unequal lengths.

Among numerous options, the following time series features have been commonly applied: autocorrelation function (ACF), partial autocorrelation function (PACF), normalized periodogram, log-normalized periodogram, cepstrum and coefficients of autoregressive models (Caiado, Crato, & Peña, 2006; D'Urso & Maharaj, 2009, 2012; Galeano & Peña, 2000; Kalpakis, Gada, & Puttagunda, 2001). We describe these time series features briefly in the following sections.

ACF and PACF

ACF and PACF have been very popular in characterizing time series. Let $\{X_t\}$ be a stationary time series with length T. Denote $\{X_{t-h}\}$ the lagged time series by h periods. The autocovariance of $\{X_t\}$ at lag h is calculated as:

$$\gamma_X(h) = Cov(X_t, X_{t-h}) = E[(X_t - \mu_X)(X_{t-h} - \mu_X)],$$

where μ_X is the expected value of $\{X_t\}$. The autocorrelation of $\{X_t\}$ at lag h is given by

$$\rho_X(h) = Cor(X_t, X_{t-h}) = \frac{\gamma_X(h)}{\gamma_X(0)} = \frac{E[(X_t - \mu_X)(X_{t-h} - \mu_X)]}{E(X_t - \mu_X)^2}.$$

The autocorrelations at lags 1, 2, ... jointly form the ACF. Its empirical version is named sample autocorrelation function (SACF). Let $x_1, x_2, \ldots, x_n$ be the observations of a time series. The sample autocovariance is computed as follows:

$$\hat{\gamma}(h) = \frac{1}{n} \sum_{t=1}^{n-|h|} (x_{t+|h|} - \bar{x})(x_t - \bar{x}), \quad -n < h < n,$$

where $\bar{x}$ is the sample average. The SACF is of the following form:

$$\hat{\rho}(h) = \frac{\hat{\gamma}(h)}{\hat{\gamma}(0)}, \quad -n < h < n.$$

If we would like to examine the relationship between current and earlier values of a time series but holding the effect of all other time lags constant, the partial autocorrelation function (PACF) should be considered. That is, the PACF measures the degree of association between $\{X_t\}$ and $\{X_{t-h}\}$, whereas the other time lags are not taken into account. The PACF of $\{X_t\}$ at lag h is a function $\alpha(\cdot)$ defined as follows:

$$\alpha(0) = 1,$$
$$\alpha(h) = \phi_{hh}, \quad h \geq 1,$$

where ϕ_{hh} is the last component of

$$\phi_h = \Gamma_h^{-1} \gamma_h,$$

where $\Gamma_h = [\gamma(i-j)]_{i,j=1,\ldots,h}$, and $\gamma_h = [\gamma(1), \ldots, \gamma(h)]'$.

Given a set of observations $\{x_1, x_2, \ldots, x_n\}$ with $x_i \neq x_j$ for some i and j, the sample PACF is expressed as

$$\hat{\alpha}(0) = 1,$$
$$\hat{\alpha}(h) = \hat{\phi}_{hh}, \quad h \geq 1,$$

where $\hat{\phi}_{hh}$ is the last component of

$$\hat{\phi}_h = \hat{\Gamma}_h^{-1} \hat{\gamma}_h.$$

Using MATLAB, the ACF and PACF of a time series realization at lag h can be computed respectively by functions "`autocorr(x, h)`" and "`parcorr(x, h)`" where "x" stands for the time series realization.

In time series analysis it is common to plot the ACF and PACF against time lags. Such plots are referred to as correlograms, which visualize time-varying patterns by serial dependence. Correlograms are very helpful in discovering cycles, seasonality or other time series patterns, as well as determining a suitable time series model for the observed data. Refer to Makridakis, Wheelwright, and Hyndman (1998) for an extensive discussion.

Normalized periodogram, log-normalized periodogram and cepstrum

The ACF and PACF are time series features defined in the time domain. In particular cases (e.g., speech recognition) it is more appropriate to consider time series features in the frequency domain. Associated with each stationary stochastic process is a spectral density function which is used to characterize frequency properties of a stationary time series. The spectral representation decomposes a stationary time series $\{X_t\}$ into a sum of sinusoidal components with uncorrelated random coefficients. In conjunction with this decomposition, there is a decomposition into sinusoids of the autocovariance function of $\{X_t\}$. For stationary time series the spectral decomposition is an analogue of the Fourier representation of deterministic functions. Spectral analysis is especially useful when analysing multivariate stationary processes and linear filters.

Let $\{X_t\}$ be a stationary time series with zero mean and autocovariance function $\gamma(\cdot)$ satisfying $\sum_{h=-\infty}^{\infty} |\gamma(h)| < \infty$. The spectral density function (also referred to as the spectrum) is the Fourier transform of the autocovariance function, namely,

$$f_X(\omega) = \frac{1}{2\pi} \left(\sum_{h=-\infty}^{\infty} \gamma(h) e^{-ih\omega} \right),$$

where $\gamma(h)$ is the autocovariance coefficient at lag h of $\{X_t\}$, and $\omega \in [-\pi, \pi]$ is the frequency. $e^{i\omega} = \cos(\omega) + i\sin(\omega)$, where $i = \sqrt{-1}$. The spectral density function is a nonnegative, even function and therefore it is sufficient to consider $\omega \in [0, \pi]$. The autocovariance of a stationary time series with absolutely summable autocovariance function can be expressed as the Fourier coefficients of the nonnegative even function:

$$\gamma(k) = \int_{-\pi}^{\pi} e^{ik\omega} f_X(\omega) d\omega = \int_{-\pi}^{\pi} \cos(k\omega) f_X(\omega) d\omega.$$

For a time series realization $\{X_t : t = 1, 2, \ldots T\}$, an estimator of the spectral density function is the periodogram. The periodogram of $\{X_t\}$ is expressed as

$$P_X(\omega_j) = \frac{1}{T} \left| \sum_{t=1}^{T} x_t e^{-it\omega_j} \right|^2,$$

where the frequencies $\omega_j = 2\pi j/T$, $j = 1, \ldots, T/2$. MATLAB has the in-built function "$[\text{px}, \text{w}] = \text{periodogram}(\text{x})$" for computing $P_X(\omega_j)$, where the output "px" and "w" stand for the periodogram power spectral density estimate and normalized frequencies, respectively.

The normalized periodogram characterizes the correlation structure, which is of the following form:

$$NP_X(\omega_j) = \frac{P_X(\omega_j)}{\hat{\gamma}_0},$$

where $\hat{\gamma}_0$ is the sample variance of time series.

Since the variance of periodogram ordinates is proportional to the spectrum value at the corresponding frequencies, it is feasible to consider the logarithm of normalized periodogram:

$$LNP_X(\omega_j) = \log NP_X(\omega_j).$$

Another commonly applied time series feature is the cepstrum, which has been proven useful in many fields of research including processing signals containing echoes seismology, measuring properties of reflecting surfaces, loudspeaker design, restoration of acoustic recordings, and detecting families of harmonics and sidebands (e.g., in gearbox and turbine). In order to define the cepstrum of a time series, the stationary autoregressive moving average (ARMA) model is considered:

$$X_t = \sum_{r=1}^{p} \phi_r X_{t-r} + \varepsilon_t + \sum_{r=1}^{q} \theta_r \varepsilon_{t-r},$$

where ϕ_r and θ_r are the autoregressive and moving average parameters, respectively, and ε_t is a white noise process whose variance is σ^2. The respective spectral density is defined as

$$f_X(\omega) = \frac{\sigma^2}{2\pi} \left| \frac{1 - \sum_{h=1}^{p} \phi_h e^{ih\omega}}{1 - \sum_{h=1}^{q} \theta_h e^{ih\omega}} \right|^2.$$

An exponential form for the log spectral density, introduced by Bloomfield (1973), is defined as

$$\lambda_X(\omega) = \log f_X(\omega) = \frac{\sigma^2}{2\pi} \exp\left(2 \sum_{h=1}^{p} \psi_h \cos(h\omega) \right),$$

where $0 < \omega < \pi$, $\psi_h (h = 1, \ldots, p)$ and σ^2 are unknown parameters. Savvides, Promponas, and Fokianos (2008) introduced an expression of the cepstrum of $\{X_t\}$, which is of the following form:

$$CP(\omega) = \log \lambda_X(\omega) = \psi_0 + 2 \sum_{h=1}^{p} \psi_h \cos(2\pi h\omega),$$

where $\psi_0 = \int_0^1 \log \lambda_X(\omega) d\omega$ is the logarithm of the variance of ε_t. If $\log \lambda_X(\omega)$ is absolutely integrable on $(0, 1)$, the cepstral coefficients are defined as

$$\psi_h = \int_0^1 \log \lambda_X(\omega) \cos(2\pi h) d\omega,$$

for $h = 0, 1, 2, \ldots$. It is worth noting that due to the convergence rate of $\log \lambda_X(\omega)$, choosing only a small number of cepstral coefficients is sufficient to describe the second order characteristics of a time series. MATLAB has the in-built function "`rceps(x)`" for the computation of cepstrum.

Autoregressive coefficients

The estimated coefficients of an autoregressive (AR) model can be employed as time series features as well.[1] Consider the following p-order AR model:

$$X_t = \phi_0 + \phi_1 X_{t-1} + \cdots + \phi_p X_{t-p} + \varepsilon_t,$$

where the error term ε_t is assumed a white noise process. In practice, the AR order p is determined by a means of model selection criterion, such as the Bayesian information criterion (BIC) proposed by Schwarz (1978). Once p is selected properly,

---

[1] In the survey of Liao (2005), these methods are categorized as model-based approaches.

the estimated values of $\phi_1, \ldots, \phi_p$ are capable of characterizing time-varying patterns of X_t. It is common in the literature to use AR coefficient estimates to classify time series realizations (Liu & Maharaj, 2013; Liu, Maharaj, & Inder, 2014; Maharaj, 2000).

Once the features of time series are computed, standard classification algorithms (e.g., k-nearest neighbour, support vector machines, etc.) can be applied. The principle of classification is straightforward: always assign a time series to the group with which it has the greatest similarity. The similarity/dissimilarity is measured by some distance function of computed time series features, for example, the Euclidean distance between SACF values. Note that in practice, time series features of a group are approximated by the features of a representative of the group (e.g., the centroid time series in that group).

There are numerous examples of classifying ECG, EMG or EEG data using feature-based methods, and the results of classification have been generally desirable. For instance, Ge et al. (2002) and Corduas and Piccolo (2008) attempted to classify ECG data by measuring the distance between estimated autoregressive coefficients. The result of classification showed that the arrhythmia groups can be well recognized and separated from the normal sinus rhythm group. Kalpakis et al. (2001) considered both ACF and cepstrum time series features for the classification of ECG signals. Both features achieved good performance in discriminating between health and illness groups. Kang, Cheng, Lai, and Tsao (1995) employed both autoregressive and cepstral coefficients to classify patterns in EMG signals of 20 repetitions of 10 motions, and experimental results showed that mean recognition rate of the cepstral coefficients was at least 5% superior to that of the autoregressive coefficients. Gupta, Parameswaran, and Lee (2009) addressed the classification issue of EEG data by characterizing electrode signals using normalized power spectral density, which resulted in very high accuracy.

In summary, feature-based methods for the classification of electrode signal data have great potential in the detection of particular illness with respect to those of healthy people. Past studies support the claim that feature-based methods are efficient and reliable in the classification of electrode signals for health monitoring purposes.

8.1.3   Fall detection

Hauer, Lamb, Jorstad, Todd, and Becker (2006) defined a fall as "an unexpected event in which the participant comes to rest on the ground, floor, or lower level". As claimed by Igual, Medrano, and Plaza (2013), falling is one of the major health risks among aged people. Approximately $28-35\%$ of people over 65 years old fall each year, while this figure increases to $32-42\%$ for those aged 70 or above. The frequency of falls increases exponentially with age-related biological changes, leading to a high incidence of falls or fall-related injuries such as soft tissue damage, fractures, superficial cuts and abrasions to the skin, etc. Since the population in most countries is aging rapidly, the number of injuries caused by falls will continue increasing if preventive measures are not taken. As a consequence, systems that are capable of detecting falls are being investigated to alleviate this problem.

According to Igual et al. (2013), a fall detection system can be defined as "an assistive device whose main objective is to alert when a fall event has occurred", which has the potential to mitigate some of the adverse consequences of a fall. For example, if a fall of an aged person is detected, then the nearest medical centre can be alerted immediately so that timely medical care becomes available to this person. This is especially important to those who lack the ability to stand up without assistance, implying another advantage of fall detection systems: the fear of falling can be reduced if an aged person is aware that his/her body movements are being monitored. Among elderly people, the fear of falling may increase the risk of suffering from a fall. Igual et al. (2013) figured out that the fear of falling tends to be associated with adverse consequences such as avoidance of activities, less physical activity, falling, depression, decreased social contact and lower quality of life. Fortunately, such problems can be solved by reliable, accurate fall detection systems. As stated by Brownsell and Hawley (2004), elderly people who are monitored by a fall detector tend to feel confident and independent, claiming that the fall detector improves their safety. It is for this reason that numerous fall detection systems have been developed over the past decade.

Intuitively, the objective of a fall detection system is straightforward: to distinguish a fall from activities of daily living (ADL) such as walking, running or jumping. That is, a fall detection system is essentially a supervised learning algorithm that recognizes a newly observed event as either a fall or an ADL. As noted by Igual et al. (2013), however, this is not an easy task to carry out, since particular ADL (sitting down, lying down, etc.) have very similar patterns to those of a fall. As a consequence, massive data need to be collected from falls and various ADL (which are either real or simulated by volunteers) to train a fall detection system.

The reliability of a fall detection system is usually evaluated by two factors: sensitivity and specificity. The former refers to the ability that a fall detector classifies a fall as a fall, while the latter denotes how likely a detector classifies an ADL as ADL. Mathematically, these two terms are defined as follows (Kangas et al., 2009):

$$Sensitivity = TP/(TP + FN) \times 100\%,$$
$$Specificity = TN/(TN + FP) \times 100\%,$$

where TP, FN, TN and FP denote true positives (detected falls), false negatives (undetected falls), true negatives (ADL samples not giving fall alarm) and false positives (ADL samples giving false fall alarm), respectively. An ideal fall detection system is able to achieve 100% in both sensitivity and specificity, that is, $FN = FP = 0$. For most fall detection systems, however, there is a trade-off between sensitivity and specificity. As stated by Chao, Chan, Tang, Chen, and Wong (2009), aiming for a higher specificity leads to fewer false alarms but may lower sensitivity, and vice versa.

Igual et al. (2013) categorize fall detection systems into two groups: context-aware systems and wearable devices. The former depends on the sensors deployed in the environment such as cameras, floor sensors, infrared sensors and pressure sensors, which are out of the scope of this chapter. Wearable devices, on the other hand, are miniature electronic sensor-based devices that are worn by a patient. The majority of these devices are based on accelerometers and gyroscopes, which have been extensively employed for gait and balance evaluation, fall risk assessment and mobility monitoring. In practice, these sensors can be attached to the body directly, or built in a smartphone. In either case, data of body movements are collected using those sensors and then analysed to determine whether a movement is a fall or an ADL. To analyse the data collected from accelerometers and gyroscopes, two types of approaches have been proposed in the literature: thresholding techniques and machine learning methods.

Thresholding techniques

Thresholding techniques, also known as threshold-based methods, are conceptually very simple: body movements during a fall exhibit features that are distinguishable from those during an ADL, which can be measured by particular quantities such as angular acceleration, change in trunk angle, etc. If these quantities exceed a certain threshold, then the corresponding event is classified as a fall.

Thresholding techniques are based on data, which are biosignals of a patient measured by wearable sensors. For the purpose of data collection, one may choose to implement one of the following three methods:

- Using accelerometers to observe acceleration data (Kangas, Konttila, Lindgren, Winblad, & Jämsä, 2008; Kangas, Konttila, Winblad, & Jämsä, 2007; Kangas et al., 2009);
- Using gyroscopes to observe angular data (Bourke & Lyons, 2008);
- Using both accelerometers and gyroscopes to observe acceleration and angular data (Li et al., 2009).

We first discuss how the acceleration data can be utilized by a thresholding technique. Kangas et al. (2009) stated that a typical fall detection system is able to identify different phases of a fall event, including (i) motion before impact, which is recognized by high velocity, posture change or free fall, (ii) impact itself, which is learnt based on high acceleration or a rapid change in acceleration, and (iii) end posture or reduced general activity after the impact. They claim that triaxial accelerometers and quite simple algorithms would be sufficient for fall detection. In the research of Kangas et al. (2007, 2008, 2009), intentional falls (forward, backward and lateral) towards a mattress were performed by healthy volunteers, while ADL samples were collected from those volunteers as well. During the falls and ADL, accelerations were measured synchronously at the waist, wrist and head with triaxial accelerometers, where each of the three axes was calibrated statically against the gravitation. The observed acceleration signals were converted into gravitational units with a custom-made MATLAB (R2006a) program.

To detect a fall, Kangas et al. (2007, 2008, 2009) considered the following parameters:

- Total sum vector. It contains both dynamic and static acceleration components, which can be computed as follows:

$$SV_{TOT} = \sqrt{A_x^2 + A_y^2 + A_z^2},$$

where A_x, A_y and A_z denote the acceleration in the x-, y-, and z-axes, respectively. When standing, SV_{TOT} is about $1g$, where g stands for the standard gravity. The start of a fall is recognized if $SV_{TOT} < 0.6g$.
- Dynamic sum vector. It is computed as

$$SV_D = \sqrt{A_{x,HP}^2 + A_{y,HP}^2 + A_{z,HP}^2},$$

where $A_{x,HP}$, $A_{y,HP}$ and $A_{z,HP}$ correspond to acceleration data that were high-pass (HP) filtered. This parameter was used to detect fall-associated impacts. When standing, SV_D is approximately $0g$.
- Sliding sum vector (SV_{MaxMin}). It was calculated using the differences between the maximum and minimum values in a 0.1-second sliding window for each axis, which can be used to investigate fast changes in the acceleration signal. When standing, SV_{MaxMin} is approximately $0g$.
- Vertical acceleration. It is of the following form:

$$Z_2 = (SV_{TOT}^2 - SV_D^2 - G^2)/2G,$$

where $G = 1g$ standing for the gravitational component. $Z_2 \approx 0$ while standing.
- Velocity (v_0). v_0 was computed by integrating SV_{TOT} over the area from the beginning of the fall until the impact.

To determine threshold values, Kangas et al. (2007) proposed to adjust parameters to achieve optimal detection of falls with minimized false alarms from ADL (maximal sensitivity with 100% specificity when possible). Kangas et al. (2009) considered the following thresholds: $2.0g$, $1.7g$, $1.5g$, $2.0g$ and 0.7 m/s for SV_{TOT}, SV_D, SV_{MaxMin}, Z_2 and v_0, respectively. Posture monitoring after a fall was also investigated. The posture was detected 2 seconds after the impact from the low-pass (LP) filtered vertical signal, based on the average acceleration in a 0.4-second time interval, with a signal value of $0.5g$ or lower considered to be a lying posture.

Using the calculated parameters and determined threshold values, Kangas et al. (2008, 2009) applied the following three algorithms to detect a fall:

- The "impact + posture" algorithm. It was based on the detection of the impact by a threshold value of SV_{TOT}, SV_D, SV_{MaxMin} or Z_2, followed by posture monitoring.
- The "start of fall + impact + posture" algorithm. The start of a fall was detected by observing that $SV_{TOT} < 0.6g$, followed by the detection of the impact within a time frame of 1 second by a threshold value of SV_{TOT} or Z_2. Posture was monitored at last.

- The "start of fall + velocity + impact + posture" algorithm. The "start of fall", "impact" and "posture" components of this algorithm were the same as those in the previous algorithm, while the "velocity" component was carried out by the detection of v_0 exceeding the threshold.

The accelerometer-based method described above showed desirable performance. Kangas et al. (2009) demonstrated that it can discriminate various types of falls from ADL, with a sensitivity of 97.5% and a specificity of 100%.

While accelerometers are able to facilitate fall detection, gyroscopes, on the other hand, have also been widely considered in the literature. The objective of a gyroscope is to measure angular velocities of a moving object, which can be used to infer angular accelerations and changes in trunk angle. Bourke and Lyons (2008) investigated a threshold-based algorithm that is capable of automatically discriminating between falls and ADL, using a biaxial gyroscope sensor. Their hypothesis was that when a person falls and hits the ground it is expected that the changes in angular acceleration, angular velocity and body angle would be different from those experienced during ADL, and therefore trunk bi-axial angular acceleration, angular velocity and body angle signals will have peak values that are distinguishable from those produced during ADL. To test this hypothesis, trunk pitch and roll readings were recorded during simulated falls and ADL by biaxial gyroscope sensors that were fitted at the sternum of participants. In their study, simulated falls towards crash mats were performed by young volunteers under supervised conditions, whereas ADL were performed by elderly subjects.

During each simulation, Bourke and Lyons (2008) recorded pitch and roll angular velocity signals, denoted ω_p and ω_r, respectively. Both ω_p and ω_r were low-pass filtered using a second-order low-pass Butterworth 2-pass digital filter with a cut-off frequency of 100 Hz. The resultant angular velocity signal, denoted ω_{res}, was derived as follows:

$$\omega_{res} = \sqrt{\omega_p^2 + \omega_r^2},$$

which provides a combined measure of the angular velocity in the sagittal and frontal planes.

As noted by Bourke and Lyons (2008), there are two possible scenarios if a thresholding technique is applied to the peak values of ω_{res}: (i) the peak values obtained from ADL do not overlap with those of a fall event, and (ii) the peak values obtained from ADL overlap with those of a fall event. Obviously, the first scenario implies a simple solution: a single threshold can be placed at the lowest peak value of falls,[2] which discriminates between falls and ADL with 100% sensitivity and specificity. In contrast, the second scenario appears much more complicated, as applying a single threshold is not sufficient to distinguish falls from ADL. To solve this problem, Bourke and Lyons (2008) proposed to investigate two

---

[2] This is under the assumption that ω_{res} values of fall events are generally higher than those of ADL.

additional measurements, namely, the resultant angular acceleration (α_{res}) and the resultant change in trunk angle (θ_{res}). Based on ω_p and ω_r, α_{res} and θ_{res} were computed as follows:

$$\alpha_{res} = \sqrt{\alpha_p^2 + \alpha_r^2},$$

$$\theta_{res} = \sqrt{\theta_p^2 + \theta_r^2},$$

where

$$\alpha_p = \frac{d}{dt}\{\omega_p\}_{-0.5s}^{+0.5s},$$

$$\alpha_r = \frac{d}{dt}\{\omega_r\}_{-0.5s}^{+0.5s},$$

$$\theta_p = \int_{t=-1.2s}^{t=+0.5s} \omega_p(t)dt,$$

$$\theta_r = \int_{t=-1.2s}^{t=+0.5s} \omega_r(t)dt,$$

and then peak values of α_{res} and θ_{res} are obtained as

$$\alpha_{max} = \max(a_{res}),$$
$$\theta_{max} = \max(\theta_{res}|_{t=0s}^{t=+0.5s}).$$

With respect to ω_{res}, α_{max} and θ_{max}, three threshold values were specified by Bourke and Lyons (2008), which are in turn $FT1 = 3.1\,\text{rads}/s$ (angular velocity threshold), $FT2 = 0.05\,\text{rads}/s^2$ (angular acceleration threshold) and $FT3 = 0.59\,\text{rad}$ (trunk angle change threshold). Note that t in the equations above denotes the time axis during a fall or ADL, with $t = 0$ referring to the time at which $FT1$ was exceeded. They then proposed the following thresholding algorithm for fall detection:

Step 1: Compute ω_{res};
Step 2: If $\omega_{res} > FT1$, proceed to *Step 3*; otherwise, no fall is detected;
Step 3: Compute α_{max} and θ_{max}. If $\alpha_{max} > FT2$ and $\theta_{max} > FT3$, then a fall is detected; otherwise, no fall is detected.

The algorithm proposed by Bourke and Lyons (2008) exhibits a hierarchy of the thresholding process, which is essentially the same as a decision tree. That is, multiple classification criteria are incorporated in a tree-like structure, where each node represents a thresholding rule. Compared to single-node classification algorithms, a tree-like thresholding structure substantially improves the reliability of the fall detection system. As demonstrated by Bourke and Lyons (2008), with 100% sensitivity ensured, a specificity of 97.5% was achieved when considering only $FT1$, whereas 100% specificity was obtained if the combination of all three thresholds was adopted.

Having witnessed that using either accelerometers or gyroscopes may lead to successful fall detection, one may wonder whether the reliability of a fall detection system can be further improved by considering both types of sensors. In the literature, attempts have been made to explore the consequences of integrating accelerometer and gyroscope sensors. Li et al. (2009) used the Technology-Enabled Medical Precision Observation (TEMPO) 3.0 sensor nodes, each of which included a triaxial accelerometer and a triaxial gyroscope. The sensor nodes were attached on the chest and thigh, given the fact that most postures have different angles between the trunk and upper legs. Three male volunteers were recruited to simulate various activities, including ADL (walk on stairs, walk, sit, jump, lay down, run, run on stairs), fall-like motions (quickly sit-down upright, quickly sit-down reclined), flat surface falls (fall forward, fall backward, fall right, fall left), inclined falls (fall on stairs). For posture monitoring, they also collected data from various static postures, such as standing, bending, sitting and lying.

Li et al. (2009) carried out fall detection in three steps: activity intensity analysis, posture analysis, and transition analysis. This algorithm can be summarized as follows:

- For a particular activity, if the change of sensor readings within a time interval did not exceed a predetermined threshold, it would be classified as a static posture; otherwise, it would be recognized as a dynamic transition. To discriminate between a dynamic transition and a static posture, readings of linear acceleration and rotational rate of the trunk and thigh were utilized, with thresholds $0.40g$ and 60 degree/second applied, respectively.
- The readings from accelerometers were then used to determine whether a posture is standing, bending, sitting or lying. To implement, they calculated the angle between the trunk and the gravitational vector (θ_A) and the angle between the thigh and the gravitational vector (θ_B). The following rules were applied:
  If $\theta_A < 35$ and $\theta_B < 35$: Standing;
  If $\theta_A > 35$ and $\theta_B < 35$: Bending;
  If $\theta_A < 35$ and $\theta_B > 35$: Sitting;
  If $\theta_A > 35$ and $\theta_B > 35$: Lying.
- Once a lying posture was identified, the previous five seconds of data were investigated to determine whether the transition to the lying posture was intentional, whereas an unintentional transition would be identified as a fall. Intentional and unintentional transitions were distinguished by applying the following thresholds to peak values of acceleration and angular rate: $3.0g$ and $2.5g$ for the acceleration recorded from the chest and thigh nodes, respectively; 200 degree/second and 340 degree/second for the angular rate recorded from the chest and thigh nodes, respectively.

Using accelerometers and gyroscopes, the fall detection algorithm of Li et al. (2009) achieved a sensitivity of 91% and a specificity of 92%. In addition, their algorithm appeared to be computationally efficient, which is advantageous to users in real-life applications.

Machine learning methods

Unlike thresholding techniques, machine learning methods do not apply a cut-off level to the observed biosignals to distinguish falls from ADL. Instead, classification algorithms are employed to achieve a decision. The following machine learning methods have been widely applied for fall detection:

- Regression models
- Naïve Bayes
- Random forest
- Decision tree
- Support vector machine
- The k-nearest neighbour classifier
- Hidden Markov models
- Artificial neural network

For details of these classification methods, the reader is referred to Chapter 2 of this book.

A typical machine learning method for fall detection is implemented as follows:

1. Collect data of falls and ADL, which could be either real or simulated;
2. Extract features from the observed data;
3. Train the machine learning algorithm using the extracted features and the corresponding labels (fall or ADL);
4. Apply the trained machine learning algorithm to make predictions.

Step 1 is very similar to the data collection process when implementing a thresholding technique, where acceleration and/or angular data are observed from wearable devices. In addition to acceleration and angular parameters that describe body movements, a machine learning method may also collect personal information features (gender, age, weight, height, etc.) in Step 2. As argued by Kerdegari, Samsudin, Ramli, and Mokaram (2012), personal information collection should be considered as a part of the feature extraction process, since personal information features have the potential to enhance the performance of a machine learning algorithm. For example, people with different heights or weights are likely to produce distinct readings of acceleration and rotational rate during a fall, while discriminating between groups of people with different physical attributes may reduce the rate of misclassification.

In Step 3, the classification algorithm is trained using the features obtained in Step 2 as well as the corresponding labels of body movements, while in Step 4 the optimized algorithm is applied to determine whether a newly observed body move-ment should be identified as a fall or an ADL. Although there is no classification algorithm that consistently outperforms the others when detecting falls, artificial neural networks tend to achieve good performance in various studies (Abbate et al., 2012; Kerdegari et al., 2012).

Limitations

It is worth noting that the fall detection systems discussed above were constructed based on the simulated fall data that were observed under constrained conditions, whereas falls in reality are generally under unconstrained conditions. This implies that statistical features of simulated falls tend to be distinct from those of real falls.

For instance, as noted by Bourke and Lyons (2008), when simulating a fall those young volunteers were instructed not to break their fall, whereas in a real situation a falling person would try to break the fall naturally, producing acceleration and angular data that are different from simulated values. In the literature, there has been evidence challenging the reliability of fall detection algorithms being applied to real data. For example, a study was carried out by Bagala et al. (2012) to evaluate fall detection algorithms based on a large dataset of real-world falls, and they concluded that algorithms that were successful at detecting simulated falls did not perform well when attempting to detect real-world falls. As a consequence, one is encouraged to determine optimal threshold values, or to train machine learning algorithms using real-world datasets, such as the one considered by Bagala et al. (2012).

8.2 Mobile devices in transportation

The benefit from using mobile devices (especially smart phones) to solve transportation problems (e.g., congestion) was emphasized by Lane et al. (2010) as well as Higuera de Frutos and Castro (2014). It is widely acknowledged that traffic remains a serious global problem in the twenty-first century, whereas the innovative utilization of mobile devices in transportation has great potential to produce solutions. Higuera de Frutos and Castro (2014) suggested that sensor-equipped mobile devices can be considered as a low-cost, efficient tool for the collection of information about road characteristics and road maintenance conditions which are essential in the management of road networks. For instance, mobile phone sensing systems can be used to provide fine-grained traffic information on a large scale using smart phones that enable accurate travel time estimation for improving commute planning. In this section, a case study is presented to demonstrate how the travel time data recorded by mobile devices are processed to gain useful information for transportation planning.

8.2.1 An overview of the case study

The primary focus of this case study is on the management of current roadways infrastructure and planning for maintenance and future growth, which are important tasks for local and state governments. The maintenance and construction of new roads are large undertakings, requiring considerable dedication of resources and potential disruption for many users. Making decisions about these activities requires a good understanding of the patterns of usage for existing roadways and a means of planning and estimating the likely impact of future additions or modification of the existing network. For such purposes, research has been undertaken extensively in the literature to estimate travel time on motorways or arterials (Bhaskar & Chung, 2013; Bhaskar et al., 2015a, Bhaskar, Qu, & Chung, 2015b; Li, Xiqun, Zhiheng, & Lei, 2013; Malinovskiy, Saunier, & Wang, 2012; Martchouk, Mannering, & Bullock, 2011; Mei, Wang, & Chen, 2012; Sun, Yang, & Mahmassani, 2008).

To monitor road use for ongoing planning and efficiency of the road network in the state of Queensland, Australia, the Queensland Department of Transport and Main Roads (TMR) collected usage data by recording the movement of media access control (MAC) addresses from Bluetooth enabled mobile devices through sensors placed at major intersections throughout Queensland. As indicated by Bhaskar et al. (2015a,b), tracking Bluetooth MAC addresses has gained much interest of researchers as one of the most cost effective ways of recording travel time. MAC IDs of discoverable Bluetooth devices being transported by road users are tracked by Bluetooth MAC scanners (BMS), and the travel time can be easily recorded by matching the MAC IDs from one BMS to another. These records are referred to as the node-to-node travel time data, which is the time difference between two BMSs at two nodes collecting identical MAC address information. According to the TMR, these data are collected from around 20% of road users, which represent a sample of the actual traffic and can be used to make inference about travel times throughout the network. Principles of Bluetooth communication and BMS data acquisition are out of the scope of this case study, and the interested reader is referred to Bhaskar and Chung (2013) for details.

The current travel time system of the TMR calculates a single travel time between nodes on the road network and delivers a single average result per period of calculation. This is carried out by computing an average of all observations after removing the outliers, where the outliers include those records below a predetermined travel speed (usually 5 km/h), those records that appear off route, and those detected by the median absolute deviation (MAD) method (see Bhaskar et al., 2015b for details of the MAD). As noted by Bhaskar et al. (2015a), however, travel time estimates obtained in this way can be biased. The primary reason is that there are multiple transport modes of road users such as cars, heavy vehicles, buses, cyclists and pedestrians, but the BMSs are not able to identify them as no information about the type of the mode and number of devices within one mode is available. Although it is evident from the literature that averaging over or taking median of node-to-node travel time data can achieve good performance when concentrating on only one type of road users (e.g., cyclists (Mei et al., 2012), pedestrians (Malinovskiy et al., 2012) or motor vehicles (Martchouk et al., 2011) on a freeway), the reliability of taking an average or a median when multiple transport modes are present is questionable. Since different transport modes are associated with potentially different patterns in travel times, data collected from different types of road users should not be treated as if they were homogeneous. In particular, different road users may travel at varying speeds in the data sample, which influence the final estimation of travel time. For instance, during peak hours cyclists and pedestrians tend to be much less influenced by congestion compared to cars or buses, while during off-peak hours cars are expected to travel faster than cyclists. As a result, it would be more reasonable to estimate travel times separately for various types of road users that utilize the entire network. To achieve this, different transport modes need to be identified from the recorded data in the first place. Since the recorded MAC addresses do not provide any information about the corresponding transport mode, the aim of this case study is to develop an algorithm to model and distinguish multiple travel patterns in node-to-node travel times, with multiple road user groups identified.

Numerous methods have been proposed in the literature for data grouping purposes, which can be categorized into supervised learning (known as classification) and unsupervised learning (known as clustering). As discussed in the previous chapters, the former is applied when the grouping of data is known, while the latter is applicable when the grouping is unknown (Liu et al., 2014). As MAC addresses do not show grouping information of road users, cluster analysis of the node-to-node travel time data is carried out in this case study.

8.2.2 Statistical methods for clustering travel times

To estimate and predict travel time for various transport modes accurately, groups of road users need to be identified in the first place. We propose to use unsupervised learning techniques to study the groupings of travel times for the following reasons:

1. MAC addresses do not indicate which transport modes are in operation during a particular period of time, and transport patterns may show large variability over time. Thus, it is not feasible to predetermine the groupings.
2. Supervised learning depends heavily on historical information, which may not always be representative in transport research since travel patterns are often influenced by external facts such as weather and incidents.

As viewed in Chapter 3, there are two types of clustering techniques, namely, crisp clustering and fuzzy clustering. Crisp clustering divides data into crisp clusters, where each individual belongs to exactly one cluster. In contrast, fuzzy clustering may determine that an item belongs to more than one cluster, producing degrees of membership that indicate the extent to which such an item belongs to those clusters. As stated by D'Urso and Maharaj (2009), crisp clustering may not be appropriate in practical situations, since in many cases there is no definite boundary between clusters and hence fuzzy clustering appears to be the better option. This claim remains valid in our study, as the boundary of travel time between different types of road users may be vague. For instance, during peak hours cyclists may travel at a speed very close to, or even higher than, that of cars, while during off-peak hours the travel time of a bus over a segment of road might be rather similar to a typical motor vehicle. In this study, both crisp and fuzzy clustering techniques are employed, and a comparison between them is carried out to indicate which one is more appropriate in studying BMS travel time data. In particular, we consider the k-means algorithm as a means of crisp clustering and the Gaussian mixture model as a means of fuzzy clustering. Both methods have been widely applied in various fields of studies (Liao, 2005). The reader is referred to Chapter 3 of this book for details of these two methods.

Since both methods are unsupervised nonhierarchical approaches, one needs to determine the number of clusters beforehand. The Silhouette coefficient, discussed in Chapter 3, is used for such purposes.

8.2.3 Data and experimental design

For demonstration purposes, we consider travel times that were recorded over two segments of road on a single day. The data used in this study were observed by the TMR on 12 November 2013 (Tuesday), recorded from multiple Bluetooth MAC address sensors which were located at the intersections on Sandgate Road with Pritchard Road, Zillmere Road and Beams Road, in north Brisbane, Queensland. The link from the Pritchard Road intersection to the Zillmere Road intersection is labelled as Link A, while the link from the Zillmere Road intersection to the Beams Road intersection is labelled as Link B. Both links are outbound from the CBD of Brisbane tracking north, and each link has two bus stops. Link A is 1.3 kilometres long with two intersections that have traffic lights in operation, while Link B is 1.1 kilometres long with one intersection that has traffic lights in operation. The speed limit is 70 km/h for both links. Two service stations and a fast food outlet are located along Link A. The locations of Bluetooth MAC address sensors, the lengths of the two links and service/food facilities are labelled on the map shown by Figure 8.1.

For Link A, a total of 3303 valid MAC addresses were scanned and matched on 12 November 2013, and hence 3303 travel times were obtained. For Link B, the sample size is 1424. Figure 8.2 displays the individual travel times recorded on 12 November 2013 for different road users on Link A and B. Peak and off-peak hours can be seen obviously from the plots. As both links are outbound from the Brisbane city, peak hours emerged in the afternoon. For Link A, two peaks can be seen from the plot. The first peak was between hours from 14:30 to 16:00, where the majority of road users took longer to travel through Link A. This is because of

Figure 8.1 Sandgate Road and its surroundings.
Source: Google Maps, https://maps.google.com.au/

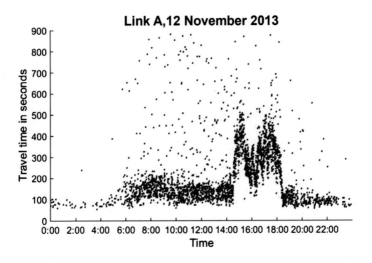

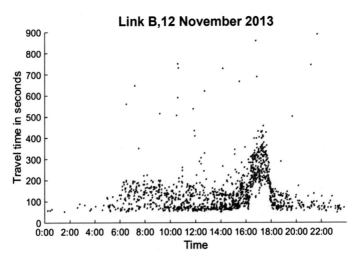

Figure 8.2 Individual travel times over Link A and B on 12 November 2013.

the school zone operating times for the Geebung Preschool and St Kevin's Catholic Primary School, which are located around 500 metres away from Sandgate road. During the afternoon school zone time period, motor vehicles picking up students from these two schools may join Sandgate road from Pritchard Road or Robinson Road East (State Route 28), causing delay in travel time over Link A. The second peak was mainly because of people getting off work in late afternoon, which started with an increase in travel time at around 16:30, reaching the maximum at around 17:00−17:30 and ending with a rapid drop at around 18:30. For Link B, 16:00−18:00 appeared to be peak hours on that day as the travel time during this period was noticeably higher than the other hours.

The clustering process using the k-means algorithm and Gaussian mixture model was carried out based on 15-min consecutive time intervals from 6:00 to 20:00. That is, travel time data recorded from 6:00 to 6:15 were clustered first, followed by those recorded from 6:15 to 6:30, and so on. Before fitting the Gaussian mixture model, the recorded travel times were transformed in natural log to approximate symmetry for each mode. For each of the 15-min time intervals, the optimal number of clusters was determined by the Silhouette coefficient and the BIC, and then clusters of data collected over the 15-min period were determined in an unsupervised manner. The determined clusters were used to estimate the travel time for various transport modes. In particular, we concentrated on the average travel time estimation for cars. During off-peak hours, cars are believed to be the quickest group of road users on average, but this is not always the case during peak hours. For instance, in congestion it is not rare to observe cyclists travelling at a faster speed than cars. As a consequence, while cars are identified as the group that has the shortest travel time during off-peak hours, we assume the largest cluster is representative of traffic conditions during peak hours and hence the average travel time of cars is estimated based on this cluster.

To justify the clustering results, two evaluations were conducted. The first evaluation aimed at comparing the estimated travel time produced by the clustering methods to some counterpart. To approximate real travel time of vehicles, the TMR also collected spot speed data of Link A, namely, data of vehicle speed collected at a single node on the link. Given the length of the link, the spot speed data were converted to travel time in seconds, and then averages were taken over consecutive 5-min time intervals from 6:00 to 20:00 as the approximations of real travel time over the day. The comparison between the results from clustering and the spot speed data can help determining if the two clustering approaches achieved reliable performance.

The second evaluation aimed at examining to what extent the produced clusters of road users are consistent over different road segments. If the travel time of a road user is repeatedly observed over two segments of a road, it is then expected that data from both segments will imply the same grouping of this road user. Consequently, a clustering method is said to be relatively consistent if a relatively high proportion of repeatedly observed road users are clustered into the same group over different segments. To carry out this evaluation, repeatedly observed road users on Link A and Link B were identified using the MAC addresses. The groupings of these road users on both links were recorded, and the proportion of consistent groupings was calculated.

8.2.4 Results and discussions

Figures 8.3 and 8.4 display the clusters determined by the two clustering methods for Link A and B, respectively. The colour of each scatter point indicates which cluster it should belong to. For the k-means algorithm, a single colour was assigned to the entire cluster as per the crisp clustering principle. For the Gaussian mixture model, the colour was assigned to each individual by its posterior probabilities

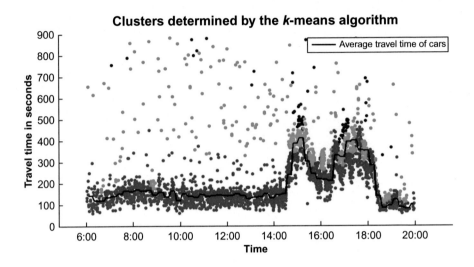

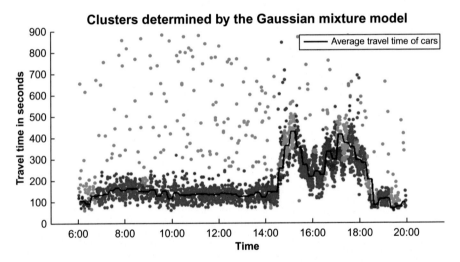

Figure 8.3 Clusters of road users and travel time estimates, Link A.

values, which coincide with the RGB colouring function in MATLAB ([1, 0, 0], [0, 1, 0] and [0, 0, 1] correspond respectively to red, green and blue). For instance, if the Gaussian mixture model determines posterior probabilities [0, 0.01, 0.99], then the colour of the corresponding scatter point is close to "pure blue"; if [0.5, 0.5, 0] are computed, the colour of that individual is then somewhere between red and green. For both clustering methods, the order of the groups was rearranged so that the colouring is consistent with the speed, that is, the red colour always represents the quickest group, with green and blue groups being slower.

In terms of the number of clusters detected, the Gaussian mixture model consistently preferred two clusters of road users for Link A, and for Link B one

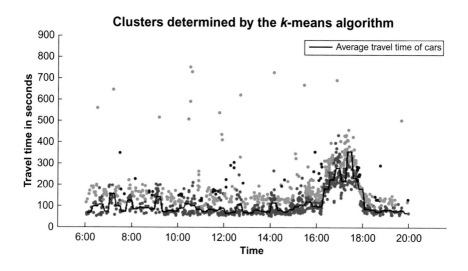

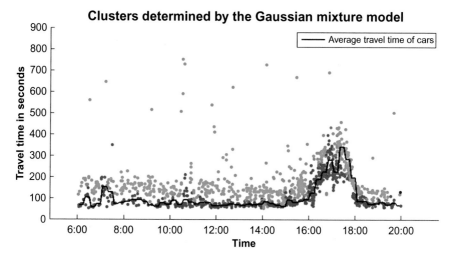

Figure 8.4 Clusters of road users and travel time estimates, Link B.

exception emerged at the last time interval (19:45−20:00) where three clusters were determined based on only a few observations. On the other hand, the k-means algorithm vacillated between 2- and 3-cluster solutions. During off-peak hours over Link A, the k-means algorithm tends to produce a larger cluster for cars compared to that from the Gaussian mixture model, resulting in slightly higher estimates of the average travel time. During peak hours, while the k-means algorithm produced distinct clusters, the fuzziness on cluster boundaries is noticeable as indicated by the Gaussian mixture model. Overall, both methods appear to track change points in traffic patterns quite well.

Figure 8.5 compares the spot speed data to the average travel time estimates obtained by the two clustering methods. It is observed that both methods were able to produce travel time estimates that follow the real traffic trend quite well over time. To examine if the clustering methods can group an individual into the same cluster on different road segments, clusters of repeatedly observed road users on Link A and B were recorded and compared. In total, travel times of 647 and 288 road users were repeatedly recorded during off-peak and peak hours, respectively. Table 8.1 reports the proportions of road users that have the same grouping over the two links. The Gaussian mixture model has higher proportions than the k-means algorithm, with the difference over off-peak period being statistically significant at 1% level of significance. This implies that the Gaussian mixture model tends to be more consistent in grouping road users, especially during off-peak hours.

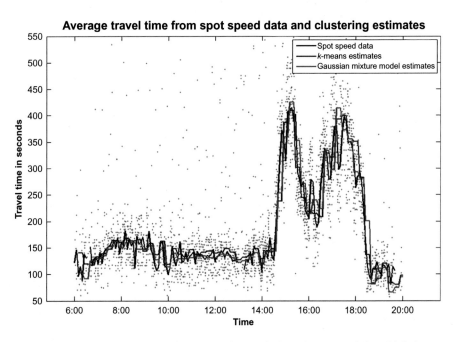

Figure 8.5 Travel time estimates of the clustering methods and spot speed data, Link A.

Table 8.1 Proportions of road users that have the same grouping over the two links

	Off-peak hours	Peak hours
k-means algorithm	57.96%	51.74%
Gaussian mixture model	75.12%	55.56%
Difference	17.16%***	3.82%

Note: *, ** and *** indicate statistical significance at the 10%, 5% and 1% level of significance, respectively.

8.2.5 Concluding remarks

Data collected from the BMSs provide information about travel patterns of individual road users, and travel time estimation can be carried out using the BMS data. To do so, many past studies take the average or the median of collected data after removing outliers, overlooking the fact that multiple transport modes are present which may have distinct travel patterns. In the literature, insufficient research has been undertaken in relation to identify multiple transport modes from recorded travel times, and we filled this gap by carrying out cluster analysis of the BMS data. The Gaussian mixture model and the k-means algorithm were employed for clustering purposes, and we carried out an empirical study on the BMS travel time data collected from segments of Sandgate Road. Both clustering methods were demonstrated to be competent in discriminating between groups of travellers, producing travel time estimates that are fairly close to the real time data. In addition, the Gaussian mixture model was believed to be more reliable in terms of determining the grouping of repeatedly observed travellers over different road segments.

The methods and results discussed in this case study provide a guideline for transport mode identification, and may contribute to further issues related to traffic monitoring such as forecasting and planning. Subsequent studies are worth carrying out to address the issue of transport monitoring further. As the TMR aims at high-quality estimation and prediction of travel time for various transport modes, methods that are capable of modelling travel time data after grouping should be explored. For instance, nonparametric smoothing techniques might be suitable to gather information about distributional properties of travel times at a specific time, while functional time series models may contribute to predicting traffic conditions.

Acknowledgement

The authors thank the Queensland Department of Transport and Main Roads (TMR) for providing the BMS travel time data. We are also grateful to Mr. Randolph Park for his valuable suggestions on Section 8.1.

References

Abbate, S., Avvenuti, M., Bonatesta, F., Cola, G., Corsini, P., & Vecchio, A. (2012). A smartphone-based fall detection system. *Pervasive and Mobile Computing, 8*, 883−899.

Aghabozorgi, S., Shirkhorshidi, A. S., & Wah, T. Y. (2015). Time-series clustering − A decade review. *Information Systems, 53*, 16−38.

Bagala, F., Becker, C., Cappello, A., Chiari, L., Aminian, K., Hausdorff, J. M., et al. (2012). Evaluation of accelerometer-based fall detection algorithms on real-world falls. *PLoS One, 7*(5). Available from http://dx.doi.org/10.1371/journal.pone.0037062.

Bhaskar, A., & Chung, E. (2013). Fundamental understanding on the use of Bluetooth scanner as a complementary transport data. *Transportation Research Part C: Emerging Technologies, 37,* 42−72.

Bhaskar, A., Kieu, L., Qu, M., Nantes, A., Miska, M., & Chung, E. (2015a). Is bus overrepresented in Bluetooth MAC scanner data? Is MAC-ID really unique? *International Journal of Intelligent Transportation Systems Research, 13,* 119−130.

Bhaskar, A., Qu, M., & Chung, E. (2015b). Bluetooth vehicle trajectories by fusing Bluetooth and loops: motorway travel time statistics. *IEEE Transactions on Intelligent Transportation Systems, 16,* 113−122.

Bloomfield, P. (1973). An exponential model for the spectrum of a scalar time series. *Biometrika, 60,* 217−226.

Bourke, A. K., & Lyons, G. M. (2008). A threshold-based fall-detection algorithm using a bi-axial gyroscope sensor. *Medical Engineering & Physics, 30,* 84−90.

Brownsell, S., & Hawley, M. S. (2004). Automatic fall detectors and the fear of falling. *Journal of Telemedicine and Telecare, 10,* 262−266.

Caiado, J., Crato, N., & Peña, D. (2006). A periodogram-based metric for time series classification. *Computational Statistics & Data Analysis, 50,* 2668−2684.

Chao, P.-K., Chan, H.-L., Tang, F.-T., Chen, Y.-C., & Wong, M.-K. (2009). A comparison of automatic fall detection by the cross-product and magnitude of tri-axial acceleration. *Physiological Measurement, 30*(10), 1027−1037.

Corduas, M., & Piccolo, D. (2008). Time series clustering and classification by the autoregressive metric. *Computational Statistics & Data Analysis, 52,* 1860−1872.

D'Urso, P., & Maharaj, E. A. (2009). Autocorrelation-based fuzzy clustering of time series. *Fuzzy Sets and Systems, 160,* 3565−3589.

D'Urso, P., & Maharaj, E. A. (2012). Wavelets-based clustering of multivariate time series. *Fuzzy Sets and Systems, 193,* 33−61.

Galeano, P., & Peña, D. (2000). Multivariate analysis in vector time series. *Resenhas, 4,* 383−404.

Ge, D., Srinivasan, N., & Krishnan, S. M. (2002). Cardiac arrhythmia classification using autoregressive modeling. *BioMedical Engineering OnLine, 1*(5), 1−12.

Gupta, A., Parameswaran, S., & Lee, C. -H. (2009). Classification of electroencephalography (EEG) signals for different mental activities using Kullback Leibler (KL) divergence. In *IEEE international conference on acoustics, speech and signal processing* (pp. 1697−1700). Taipei: IEEE. http://dx.doi.org/10.1109/ICASSP.2009.4959929.

Hauer, K., Lamb, S. E., Jorstad, E. C., Todd, C., & Becker, C. (2006). Systematic review of definitions and methods of measuring falls in randomised controlled fall prevention trials. *Age and Ageing, 35,* 5−10.

Higuera de Frutos, S., & Castro, M. (2014). Using smartphones as a very low-cost tool for road inventories. *Transportation Research Part C: Emerging Technologies, 38,* 136−145.

Igual, R., Medrano, C., & Plaza, I. (2013). Challenges, issues and trends in fall detection systems. *BioMedical Engineering OnLine, 12*(66). Available from http://dx.doi.org/ 10.1186/1475-925X-12-66.

Kalpakis, K., Gada, D., & Puttagunda, V. (2001). Distance measures for effective clustering of ARIMA time series. In *Proceedings of IEEE international conference on data mining* (pp. 273−280). San Jose: IEEE.

Kang, W., Cheng, C., Lai, J., & Tsao, H. (1995). The application of cepstral coefficients and maximum likelihood method in EGM pattern recognition. *IEEE Transactions on Biomedical Engineering, 42,* 777−785.

Kangas, M., Konttila, A., Lindgren, P., Winblad, I., & Jämsä, T. (2008). Comparison of low-complexity fall detection algorithms for body attached accelerometers. *Gait & Posture, 28*(2), 285−291.

Kangas, M., Konttila, A., Winblad, I., & Jämsä, T. (2007). Determination of simple thresholds for accelerometry-based parameters for fall detection. In *Proceedings of the 29th annual international conference of the IEEE EMBS*(1) (pp. 1367−1370). Lyon: IEEE.

Kangas, M., Vikman, I., Wiklander, J., Lindgren, P., Nyberg, L., & Jamsa, T. (2009). Sensitivity and specificity of fall detection in people aged 40 years and over. *Gait & Posture, 29*, 571−574.

Kerdegari, H., Samsudin, K., Ramli, A. R., & Mokaram, S. (2012). Evaluation of fall detection classification approaches. In *Proceedings of the fourth international conference on intelligent and advanced systems* (pp. 131−136). Kuala Lumpur: IEEE. http://dx.doi.org/10.1109/ICIAS.2012.6306174.

Lane, N. D., Miluzzo, E., Lu, H., Peebles, D., Choudhury, T., Campbell, A. T., et al. (2010). A survey of mobile phone sensing. *IEEE Communications Magazine, 48*, 140−150.

Li, L., Xiqun, C., Zhiheng, L., & Lei, Z. (2013). Freeway travel-time estimation based on temporal & spatial queueing model. *IEEE Transactions on Intelligent Transportation Systems, 14*, 1536−1541.

Li, Q., Stankovic, J. A., Hanson, M., Barth, A., Lach, J., & Zhou, G. (2009). Accurate, fast fall detection using gyroscopes and accelerometer-derived posture information. In *Proceedings of the sixth international workshop on wearable and implantable body sensor networks* (pp. 138−143). Berkeley, CA: IEEE.

Liao, T. W. (2005). Clustering of time series data − A survey. *Pattern Recognition, 38*, 1857−1874.

Liu, S., & Maharaj, E. A. (2013). A hypothesis test using bias-adjusted AR estimators for classifying time series in small samples. *Computational Statistics & Data Analysis, 60*, 32−49.

Liu, S., Maharaj, E. A., & Inder, B. (2014). Polarization of forecast densities: A new approach to time series classification. *Computational Statistics & Data Analysis, 70*, 345−361.

Maharaj, E. A. (2000). Clusters of time series. *Journal of Classification, 17*, 297−314.

Maharaj, E. A. (2014). Classification of cyclical time series using complex demodulation. *Statistics and Computing, 24*, 1031−1046.

Makridakis, S., Wheelwright, S. C., & Hyndman, R. J. (1998). *Forecasting: Methods and applications* (3rd ed.). John Wiley & Sons, Inc..

Malinovskiy, Y., Saunier, N., & Wang, Y. (2012). Analysis of pedestrian travel with static Bluetooth sensors. *Transportation Research Record, 2299*, 137−149.

Martchouk, M., Mannering, F., & Bullock, D. (2011). Analysis of freeway travel time variability using Bluetooth detection. *Journal of Transportation Engineering, 137*, 697−704.

Mattioli, F. E. R., Lamounier, E. A., Jr., Cardoso A., Soares, A. B., & Andrade, A. O. (2011). Classification of EMG signals using artificial neural networks for virtual hand prosthesis control. In *Proceedings of IEEE conference on engineering in medicine and biology society* (pp. 7254−7257). Boston, MA: IEEE. http://dx.doi.org/10.1109/IEMBS.2011.6091833.

Mei, Z., Wang, D., & Chen, J. (2012). Investigation with Bluetooth sensors of bicycle travel time estimation on a short corridor. *International Journal of Distributed Sensor Networks*. Available from http://dx.doi.org/10.1155/2012/303521.

Pantelopoulos, A., & Bourbakis, N. G. (2010). A survey on wearable sensor-based systems for health monitoring and prognosis. *IEEE Transactions on Systems, Man, and Cybernetics, Part C: Applications and Reviews*, *40*, 1–12.

Patel, S., Park, H., Bonato, P., Chan, L., & Rodgers, M. (2012). A review of wearable sensors and systems with application in rehabilitation. *Journal of NeuroEngineering and Rehabilitation*, *9*, 21.

Savvides, A., Promponas, V. J., & Fokianos, K. (2008). Clustering of biological time series by cepstral coefficients based distances. *Pattern Recognition*, *41*, 2398–2412.

Schwarz, G. (1978). Estimating the dimension of a model. *Annals of Statistics*, *6*, 461–464.

Sun, L., Yang, J., & Mahmassani, H. (2008). Travel time estimation based on piecewise truncated quadratic speed trajectory. *Transportation Research Part A: Policy and Practice*, *42*, 173–186.

Conclusion

As we have seen in this book, big data is revolutionizing our lives. We are lucky to be witnessing this revolution, which will bring significant changes to the way that the world functions in the near future. Like other scientific developments, big data is another step forward in the history of human civilization. This leads to one last question: how big would this step be?

Well, let us make the following conjecture:

Ultimately, we will be able to observe everything, explain everything and predict everything.

Big data is essentially a process of knowledge discovery through a full view of a complex system rather than isolated views. Consequently, it will enable random events to become more predictable and therefore we will be able to understand more about how and why things happen. In fact, we assume an event is unpredictable only because we are not able to collect sufficient information and hence the mechanism behind the event is not thoroughly investigated. For instance, when flipping a coin people tend to think the outcome is unpredictable because they do not have sufficient information to determine whether it will be heads or tail. However, at the moment the coin is released, the outcome is actually predetermined by the laws of physics as well as numerous features of the coin such as launch position, weight, angular velocity, the coefficient of friction, the coefficient of restitution, etc. If someone were aware of all the features influencing the coin toss and could make extremely fast calculations, they would be able to predict the outcome with 100% accuracy before the coin hits the ground.

Thanks to big data, the amount of information being recorded and analysed is exploding. In particular, things like medicine, education, advertising, news, media and entertainment are becoming more and more personalized to individuals, which offers great potential in terms of efficiency in communication and information transfer. As a result, it is not surprising to see that big data is becoming an expert in almost all fields. Did you know that social media knows you better than your relatives, friends, even yourself? Did you know that the state-of-the-art facial recognition system is even more reliable than human eyes? Did you know that your web search engine makes more accurate predictions about influenza epidemics than your general practitioner? Whereas many people think the major scientific discoveries have already happened, we believe that big data will have its say.

Index

Printed in the United States
By Bookmasters